T0332518

ELECTRONIC CAD FRAMEWORKS

THE KLUWER INTERNATIONAL SERIES
IN ENGINEERING AND COMPUTER SCIENCE

VLSI, COMPUTER ARCHITECTURE AND
DIGITAL SIGNAL PROCESSING
Consulting Editor
Jonathan Allen

Latest Titles

ELECTRONIC CAD FRAMEWORKS

by

Timothy J. Barnes
Cadence Design Systems

David Harrison
A. Richard Newton
Rick L. Spickelmier
University of California
Berkeley

Kluwer Academic Publishers
Boston/Dordrecht/London

Distributors for North America:
Kluwer Academic Publishers
101 Philip Drive
Assinippi Park
Norwell, Massachusetts 02061 USA

Distributors for all other countries:
Kluwer Academic Publishers Group
Distribution Centre
Post Office Box 322
3300 AH Dordrecht, THE NETHERLANDS

Library of Congress Cataloging-in-Publication Data

Electronic CAD frameworks / by Timothy J. Barnes . . . [et al.].
     p. cm. -- (The Kluwer international series in engineering and
computer science. VLSI, computer architecture, and digital signal
processing)
     Includes bibliographical references.
     ISBN 0-7923-9252-3
     1. Computer-aided design.   2. Engineering design-Data processing.
I. Barnes, Timothy J.   II. Series.
TA174.E39   1992
620'.0042'0285--dc20   92-10976
                                                CIP

Printed on acid-free paper.

Printed in the United States of America

TABLE OF CONTENTS

LIST OF FIGURES

FOREWORD

When it comes to frameworks, the familiar story of the elephant and the six blind philosophers seems to apply. As each philosopher encountered a separate part of the elephant, each pronounced his considered, but flawed judgement. One blind philosopher felt a leg and thought it a tree. Another felt the tail and thought he held a rope. Another felt the elephant's flank and thought he stood before a wall.

We're supposed to learn about snap judgements from this allegory, but its author might well have been describing design automation frameworks. For in the reality of today's product development requirements, a framework must be many things to many people.

As the authors of this book note, framework design is an optimization problem. Somehow, it has to be both a superior rope for one and a tremendous tree for another. Somehow it needs to provide a standard environment for exploiting the full potential of computer-aided engineering tools. And, somehow, it has to make real such abstractions as interoperability and interchangeability.

For years, we've talked about a framework as something that provides application-oriented services, just as an operating system provides system-level support. And for years, that simple statement has hid the tremendous complexity of actually providing those services.

Until this book, the knowledge of just how a framework will actually accomplish its goals has been scattered - in professional papers, in documents of the CAD Framework Initiative, in the minds of industry experts such as the authors themselves. With this book, however, anyone whose work is (or will be) touched by frameworks will gain an appreciation of the breadth and depth of the problem - and, most important, the methods of its solution.

And the authors also provide us with some very important lessons. One of the most important is the authors' warning against Brooks' "second system syndrome," where we sometimes attempt too much the second time around, because we've learned just enough from the first experience to be dangerous. As CFI framework standards diffuse through the industry, this is an apt warning for developers and users against expecting too much too soon.

But there is one thing we can expect. Today's emphasis on Electronic CAD Frameworks is only the beginning. Electronic CAD Frameworks are a pragmatic response to a burning industry need. But they're also a reflection of the industry's broader desires: to gain better leverage from its investment in design automation tools and systems - for all aspects of product development, including electronics design as well as mechanical design and software design.

If standards efforts such as those in progress within CFI point the way to this grander vision, it will be possible because of the broader appreciation of framework technologies afforded by books such as this.

Andy Graham
President
CAD Framework Initiative, Inc.

ELECTRONIC CAD FRAMEWORKS

1 INTRODUCTION

It has now been over a quarter century since the computer was introduced as an important tool in the design of integrated circuits (IC's) and systems and over the past decade it has become indispensable. Along with the rapid growth in the complexity of integrated circuits and digital systems has come an even more rapid growth in the complexity of the software tools and associated data needed to represent a design. A typical Computer-Aided Design (CAD) or Computer-Aided Engineering (CAE) system today consists of over one million lines of source code and many CAD systems contain over ten million lines of code. The data needed to describe a state-of-the-art integrated circuit, representing about 500,000 equivalent logic gates or a million transistors, can exceed two gigabytes. While the CAD tools themselves are essential to the design process, the management of such vast amounts of data and its presentation, in a useful and efficient form, to CAD programs, to designers, and to manufacturing equipment alike, has become a major issue in the electronics industry.

1.1   The Nature of a CAD Framework

The term *CAD Framework* has come to mean all of the underlying facilities provided to the CAD tool *developer*, the CAD system *integrator*, and the *end user* (IC or system designer) which are necessary to facilitate their tasks.

Broadly speaking, these three groups of people represent the users of the CAD Framework, each with their own needs and particular emphasis. The CAD Framework plays an analogous role in the development of engineering-specific, or even electrical-engineering-specific, software systems to the role played by an operating system for the development of general-purpose software applications, or the role of a specific programming environment for software development in a particular programming language. It represents a collection of *mechanisms* or *facilities* (programming libraries, extension languages, data management facilities, user interface facilities, etc.), at many different levels of abstraction, that are, to varying degrees, specific to the electronic CAD world. The use of those mechanisms to develop a particular CAD system, optimized for a specific set of end-user needs, is then the task of the tool developers, tool integrators, and often the end users themselves. For example, a Framework might be used to configure a set of tools and to develop appropriate interfaces to support schematic capture, simulation, timing verification and test generation for gate arrays; or symbolic layout editing, layout compaction, verification, and mask-pattern generation for custom CMOS; or even entry of a behavioral description, design partitioning into multiple chips, and design synthesis for a family of digital signal processors (DSPs).

The major test of a CAD Framework is that it reduce the time and cost needed to develop or modify a CAD system such that it meets the needs of its end-users. Unfortunately, this seemingly simple test represents a very broad and difficult set of requirements, many of which represent interacting trade-offs, and most of which are evolving rapidly with time, as described later.

Another important feature to note about both the definition of a CAD Framework and the test for success is that neither definition includes a particular set of features or architectural requirements. That is, there is no mention of such things as database, editors, data representation, tool flow control or the like in either definition. The particular way one satisfies the test presented above has also evolved with time and will continue to do so as we learn more about the design process, as both the software and hardware architectures of computer systems evolve, and as the needs and priorities of the end-users change. In other words, there will never be a "right answer" to the engineering Framework problem, only good answers and better answers! We believe that a clear understanding of this fact has a very important influence on the approach one takes to the design of a CAD Framework, and that those who believe they know the "right answer," as many have believed in the past, are doomed to failure. By the time they have finished their development, either the needs of the end-user have evolved or the developers have chosen an unacceptable set of trade-offs (e.g. performance versus memory requirements or flexibility). The most successful Frameworks developed to date have been designed with flexibility and easy modification as a key goal, from the early file and translator-based systems of a decade ago to the more sophisticated Frameworks of today.

1.2   The Evolution of CAD Frameworks

In the 1960's, design data management and user interface was not a major issue for IC design - the entire database often consisted of a box of punched cards and a hand-drawn roll of mylar that the designer carried to the mask shop. In the early and mid 1970's, as circuit complexity increased, proprietary and tool-dependent textual or binary data formats were developed to represent particular classes of design data, such as mask layout data (e.g. [CAL81]) and transistor or gate-level netlist descriptions(e.g. [SZY76, NAG75]). Since most CAD programs were developed independent of one another and had their own input formats, coupling them together to form an integrated system for IC design involved writing translators to and from each program. In the worst case, for n programs, $(n - 1)n$ translators would be needed, as illustrated in Figure 1.1. However, as the CAD tools evolved, their input and output formats changed along with them. As a result, it was often necessary to keep a family of translators for each program, with each translator corresponding to a different version of the input data format. Maintaining such a family of translators soon became a CAD

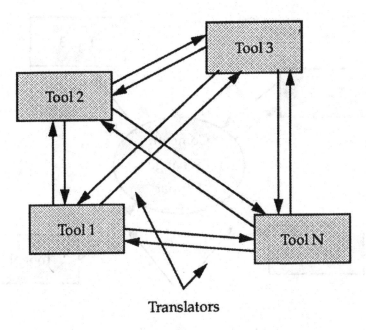

Translators

Figure 1.1: Inter-tool data translation in the absence of standards

manager's nightmare! The number of translators can be reduced to a worst-case of *2n* by choosing a common, central format and translating to and from that format, as shown in Figure 1.2. A number of *de facto* standard formats evolved in the late 1970's to meet the need for a common format and different companies standardized internally on one format for each class of data. In the mid and late 1970's, a number of public-domain standard formats were adopted of which the most successful examples are the CIF (Caltech Intermediate Form) [MEA79] for mask-level

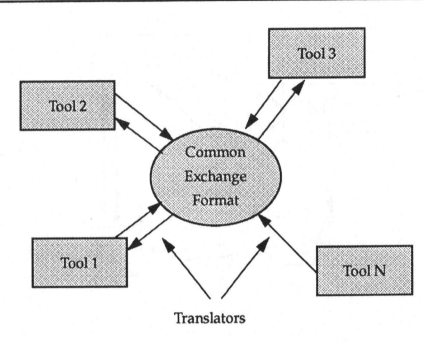

Translators

Figure 1.2: A common exchange format simplifies the translation problem

layout descriptions and SDL (Stanford Design Language) [SDL76] for gate-level netlist descriptions. While such formats provide a consistent way of storing the design data, they provide no support for managing the data. Which copy is the latest version? Has the layout been changed since the schematic diagram was updated? If I change this cell, which cells that use it will be affected? It is the ability to answer such queries that differentiates a true data management system from a simple data repository. The data representation question - how the semantics of the design information is represented in the computer - is addressed in Chapter 3, while the data management aspects of CAD Frameworks are discussed in Chapter 4.

Collections of files and translators, where each CAD tool had its own user interface, operated autonomously, and read and wrote it's own file formats, formed the first primitive CAD Frameworks. Though one can argue the effectiveness of such an approach, at the time such systems were the only way to link tools and users on a common design problem. In fact, this general approach is still in use in the majority of circuit and system design companies today. Though many end users rely on turn-key vendor-supplied CAD systems today, even some of these vendor-supplied systems still use this loosely-coupled approach. Over the years, a number of facilities have been added to improve the quality, data-management, and tool-flow aspects of such systems and an excellent early example of such work is the Designer's Workbench [ONE79].

In parallel with this work, a number of companies developed conventional database systems for managing their IC design data. Since the CAD tools all had their own user interface, only shared common operating-system resources at the level of a text-based terminal or vector plot package, and since most operating systems only supported a single interactive task per user, it is understandable that early work on Frameworks focussed on the data management aspect.

Often the companies using conventional databases for IC design were the large computer or system houses who had experience with the use of database management techniques for discrete digital system design. These record-oriented database management systems (DBMS) were developed to manage IC parts inventories, part location on standard printed circuit (PC) boards, and the connections among IC pins necessary to implement the logic schematic. These lists of connections, used to guide wire-wrap or stitch-weld machines, are generally referred to as netlists. While these companies found that the application of conventional relational, network, or hierarchical database management techniques [ULL82] was effective for structured, semi-custom design styles like gate-array and standard-cell, these approaches were not successful for custom design styles or in situations where the underlying process technology and design style was evolving rapidly [ROS80, WEI80]. Some of the specific reasons why these systems are not sufficient, even today, are described in more detail in Chapter 3.

The same rapid increases in complexity that makes the use of conventional database management techniques difficult has made the need for a unified data management system critical, especially for full-custom or structured-custom design styles. No longer is the entire design process the responsibility of a small, tight-knit group but rather teams of system designers, logic designers, circuit designers, and layout technicians must all work together and share the vast amount of data representing a modern IC-based system.

The representations of IC design data, such as mask layout, schematic diagrams, documentation, simulator input and output, are quite diverse and new data representations for this and other information are being developed continuously. This evolution requires a flexible data management system which can adapt readily to new design methods, while maintaining acceptable performance, and such a facility is a key component of an engineering Framework. The state of developments for engineering design data management is presented in Chapter 4. In the past decade, the notion of procedural circuit design [JOH78, BAT80] and rule-based expert system technology have emerged as key components in the engineering design process. These paradigms have broadened the requirements for an engineering data management system. Each object in a design may be described by data, such as its mask layout, by a local procedure, such as a parameterized cell-generator description, by generic synthesis tools, such as a channel router or a placement program, or by a combination of all of these techniques. This issue is presented in more detail in Chapter 3 and Chapter 4.

1.3 CAD Frameworks and Openness

The topic of CAD Frameworks has received a great deal of attention in the past few years, motivated largely by end-users. As designs become more complex, the design data represents the "life blood" of an IC or electronic system design company. If a particular design tool does not function correctly under certain conditions, or a workstation or mainframe computer fails, the problem can generally be overcome and work can proceed. However, if the design data were to be lost in the middle of a large design project, the cost could be astronomical. Not only would the investment in design effort be lost to that point but such a situation would also cost valuable time and a market window might be missed. This is one reason why most IC design companies have resisted trusting all of their data management tasks to a single vendor, particularly if it is not possible to archive all of the data in a non-proprietary format. In addition, once a company has committed their data to a particular vendor's system, they are "locked in" to that vendor unless there is a way of migrating the data to another system. Until recently, many system houses have used proprietary CAD systems to augment their internal efforts. That is, the turnkey system is used as a front-end for certain aspects of the design, the data is then transferred, usually via a common textual format, into the corporate design system for final checks and transfer to manufacturing. By using published textual data formats to represent the design information at certain stages, CAD vendors like Dazix, Mentor Graphics, and Viewlogic have provided a form of loosely-coupled openness that has met the needs of the end user in many situations. Support for industry-standard data formats such as EDIF [EIA87] and, more recently, VHDL [INT85] have enhanced this capability.

1.4 The Rise of Commercial CAD Frameworks

The first vendor to use the term *Framework* in product marketing and to provide a more tightly-coupled integration of tools was SDA Systems, now called Cadence Design Systems. Cadence's Framework products, along with their extension language SKILL™ [BAR90] have found widespread use in the IC design market segment, and are increasingly seen in the systems business. More recently, EDA Systems Inc. now owned by Digital Equipment Corporation) has developed a general-purpose CAD Framework [BRO87], with emphasis on the tasks of integrating "foreign" tools into a single CAD system and managing the history of the design data. The EDA product played a major role in popularizing the Framework concept in the electronic CAD community. More recently Viewlogic, Mentor Graphics and Dazix have announced Framework products.

1.5 The Impact of Object-Oriented Techniques

As software systems continue to grow in size and complexity, programmers have turned to object-oriented approaches to code development and support (for example [MOO86, TES81, STE83,

MEY89, STR78]). The next generation of workstations, with an order of magnitude increase in performance at the desk top for comparable price to workstations of today, will be a key factor in making such approaches practical and affordable outside the research laboratory. In an analogous way, IC designers are already using procedural descriptions of design components, akin to the objects in many of these languages. In addition, the database management community is directing its attention to the management of object-based descriptions of systems. From an IC design point of view, these three technologies are converging in the next generation of data management and programming systems for IC design. We expect that the interfaces to these systems will be indistinguishable from that of an object-oriented, message-based programming environment. A number of new companies have been formed to address the issue of object-oriented data management for engineering applications and these include Itasca, Object Design, Objectivity, Ontologic, Serviologic and Versant. One of the major challenges these companies face is that of adding the engineering data management-specific and design management-specific features to their environments, while also meeting the very high performance requirements of the engineering world.

1.6   The Standardization of CAD Frameworks

With such a long history of development, many researchers consider certain aspects of a CAD Framework to be understood well enough that effective standards can be established. In parallel with the ongoing Framework research efforts, a number of groups have begun work to standardize some of the interfaces to a CAD Framework. This activity was pioneered by the Engineering Information System (EIS) [LIN86a, LIN86b] project of the US Department of Defense, while the most important ongoing effort today is that of the CAD Framework Initiative (CFI), formed by an international group of companies and university participants. The stated mission of the CFI is to "develop industry acceptable guidelines for Design Automation Frameworks which will enable the coexistence and cooperation of a variety of tools", and they have already been able to demonstrate significant progress, especially in the area of design representation.

There is also considerable interest in the development of Framework standards in the European community, based on a number of significant ongoing research activities. In particular, the work at NMP-Cadlab [MIL89, GOT87], Delft University [WID88, WOL88, DEW86], and the Ireen system developed by Piloty et al [PIL89] has been a major factor in these developments. An active standardization effort in Europe began some years ago under

European "CAD Integration Project (ECIP) sponsorship, and has continued under this and other joint industry-government programs. In addition, European researchers are playing a particularly active role in the CFI developments.

The most significant challenges faced by such a group include choosing the right levels for standardization so as not to preclude further important research and development in the future, and the establishment of a forum for evolving appropriate standard data representations, in terms of their data model, for electronic CAD information. These groups are, however, also addressing the needs of a CAD Framework in the areas of user interface, inter-tool communication, portability and methodology management as well.

1.7   Organization of the Book

After presenting a general discussion of the major components of a CAD Framework mentioned above and their relationships to one another (Chapter 2: MAJOR COMPONENTS OF AN ENGINEERING FRAMEWORK on page 17), each area is presented in more detail. A brief review of the state-of-the-art and current directions for research are presented. Since the approach taken to the development and enhancement of CAD Frameworks has had more impact on their success or failure than any particular design decision, some observations on this topic are included in

Chapter 9: IMPLEMENTING A CAD FRAMEWORK on page
139. Finally, the influence of related disciplines and the ongoing
software standards efforts on the area of CAD environments is
reviewed.

2 MAJOR COMPONENTS OF AN ENGINEERING FRAMEWORK

2.1 Overview

A coarse view of the major components of an engineering Framework today is shown in Figure 2.1. A well-designed Framework provides many layers of abstraction and, in the most successful examples, all of these layers are provided to the tool developers and CAD systems integrators for their use. That is, the systems integrator can choose to use high-level facilities provided by the Framework, lower-level facilities, or even system calls provided by the operating system itself, if necessary. This is analogous to a systems programmer using a high-level language for some parts of a program but assembly code for certain critical parts. Of

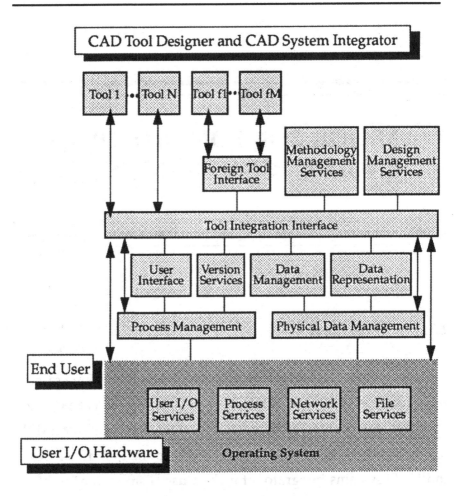

Figure 2.1: Framework Components

course, a good systems programmer only uses assembly language when there is no other way of meeting the design goals of the system.

2.2 Operating System Services

The Framework is built on the existing services provided by the operating system, which include facilities for manipulating and organizing files (*File Services*), running programs (*Process Services*), communication with other computers via electronic networks (*Network Services*) and communication with the human users of the computer system (*User I/O Services*). These services are usually provided to programs via one or more libraries of subroutine calls and to human users via a textual interface, often called a *shell*.

Since not all operating systems provide the same services, these facilities are often converted to a single abstract operating system view that can be implemented on many underlying operating systems but which presents a common interface to the Framework code itself. This interface involves the major components of physical data management and process management. Physical data management refers to all of the tasks having to do with the management of raw data either on the host machine or stored elsewhere on the network and accessible to the Framework. Process management refers to the facilities needed to run computer programs (mostly CAD tools in this case) either on the local machine or on other machines on the network.

2.3   Tool Integration and Encapsulation

The facilities that form the tool integration environment itself, as seen by the tool developer and the CAD system integrator, also include higher-level facilities for constructing user interfaces (*User Interface Services*), managing the CAD data associated with the design and coordinating access to the data by multiple CAD tools or human users (*Data Management Services*), managing the evolution, or history, of the design (*Version Services*), and facilities for defining the legal organization of the data and what particular data items, and their relationships to other data items, represent (*Data Representation Services*). Each of these aspects of the Framework are described in more detail below.

Unfortunately, CAD system integrators must live with existing design styles and tools, many of which are not available in source code form for proprietary or historical reasons. In that case, the system integrator is forced to encapsulate the tool in such a way that the tool "sees" the input and output file formats that it expects, while the data is actually being managed by the Framework. The software that implements this encapsulation is known as a *Foreign Tool Interface*. Sometimes, the foreign tool interface actually translates the information to and from the common data representation provided by the Framework, but in many cases in simply treats the entire foreign tool's input or output data as a single data record, storing it in the native format of the tool. Data manipulated in this way is treated as a single, coarse-grained "chunk" by the Framework, and is sometimes referred to as *stranger* data. In Figure 2.1, CAD tools *1 - n* represent "native" tools that are tightly integrated with the Framework at a procedural level while tools *f1 -f m* represent "foreign" tools.

2.4   Design Management

In addition to the facilities provided for tool integration, many CAD Frameworks under development today provide a variety of *meta* services - that is, services that use the integration interface themselves to provide higher-level or peripheral help with the design. For example, design management services, or design methodology services, might be provided to help the systems integrator or (more likely) the end user to specify certain "recipes" for design that may involve the sequential or concurrent execution of many tools on perhaps many different parts of the design database. Once an end user has determined that a particular combination of tools is what is required to perform a particular design task, e.g. standard-cell placement and routing or automatic design-rule checking and connectivity verification, then the flow of tools, along with their input and output needs, can be captured in a form that can be re-run automatically, as needed and for new designs. Project management services might include tools for evaluating and displaying the progress of a design.

In Figure 2.1, the end user is shown accessing the system via the operating-system user input-output services to stress the fact that the CAD Framework, like any other program operating on the computer, must communicate with the outside world via the operating system. As mentioned earlier, the area of user interface has not progressed as rapidly as some other aspects of the Framework. Until quite recently, a tool developer could not assume much more than a relatively simple textual input/output device was available for communicating with the user, and all tools made the assumption that while they were running no other tool would be using the terminal. In the case of graphics-intensive applications, the CAD programs were generally opti-mized for specific graphics hardware which, again, "owned" the display. This led to a proliferation of unique, hard-wired inter-faces for each tool and there has been very little sharing of user-interface facilities to date, beyond the use of common plotting libraries in some cases. With the advent of bit-mapped, high-resolution workstations, some attempts are being made to stan-dardize user interface facilities and to provide standard features at higher levels of abstraction as well [SCH86]. However, this work has only captured widespread attention in the engineering community in the past two years.

We believe that the area of user interface and the usability of CAD systems is key to the success of future CAD Frameworks, if engineering design systems are to find their way into the broad base of potential users. Engineers and scientists who are experts in their particular disciplines should be able to work with a system that speaks the language of their discipline. They should not have to learn the often esoteric idiosyncrasies of general-purpose computer operating systems, like Unix or DOS, let alone the features of a proprietary computer system, to get their job done. If the CAD tool developer and CAD system integrator are to be able to provide such domain-specific interfaces at reasonable cost, in reasonable time, and such that they can be ported to new generations of hardware platform, the CAD Framework must provide a wide range of facilities for developing such interfaces and these facilities must be based on the evolving standards. These issues are presented in more detail in Chapter 7.

Even with an efficient data management system and a state-of-the-art user interface, if the meaning of the numerous bytes stored and retrieved by the data manager is known only to a single tool, then the goal of integrating a number of tools to work on a common design problem and to share data cannot be met. For this reason, a common representation for the information associated with the design must be established so that two tools which read the same data interpret it the same way. This task is usually divided into the phases of agreeing on a mechanism for representing the information - a common way of representing data items and the relationships between data items - and then a common meaning for a particular set of data items and their relationships. This issue has been studied extensively both in the database area and also the programming semantics area. Unfortunately, a formal semantic model for the representation of data in electronic design, though possible, is not of much practical use today[1]. Rather, an "axiomatic" approach is taken where groups of tool developers meet and, after much discussion, agree on a particular way of representing design information and its meaning. The process is much like learning a language, where the meaning of words is learned by example. In addition, the rules for composing grammatically-correct sentences from words can be formalized, but the actual meaning of a sentence generally transcends a straightforward, bottom-up analysis of the sentence and its components. The mechanisms used for

---

1. While such models are not practical for mask, PC board, or electrical levels of description, they are of value for discrete-valued levels of description - logic gate, register, and behavioral. The use of formal data models in these areas is a topic of active research at this time.

representing information in a CAD Framework, and some examples of the ways in which they are used, are presented in Chapter 3.

With the facilities provided above, a tool integrator should be able build a CAD system and an associated set of user interfaces such that the end-user can complete a design. Unfortunately, for the complex designs of today, there are a number of important meta-issues that remain. Most design tasks today are not treated with a single, "flat" organization. Rather, the design is divided into sub-modules or sub-cells which are designed separately and in parallel. The sub-cells may themselves be composed of sub-cells, to form a hierarchy. During the evolution of the design over time, different versions of each sub-cell may be developed - the first version, the improved version, the final version, etc. With the design of the components of a system being developed in parallel, often by different designers, on different computer systems or even at different locations, it is important that when the final design is assembled, all the right components come together. Facilities for the management of the history of the design - versions and alternatives for each component, as well as specific configurations of collections of components - must be provided by the Framework. These issues are described in more detail in Chapter 4.

While developing a particular design style or design method, it is not unusual for the Framework user to execute each tool manually and in a specific order to complete the entire design or the design of a sub-cell. In this way, the user can experiment with different approaches to the design and correct errors or omissions in the tools and the design representation. Once a particular design flow, or tool flow, has been debugged, it is often useful to record and encapsulate the flow for use by others. In this way, the collection of tools and their operation as a group appear as a "super tool" to the user. The facilities necessary to support this activity fall into a category referred to as design flow management or methodology management, and are reviewed in Chapter 6.

As mentioned earlier, the boundaries of what constitutes a Framework and what does not are not defined in terms of specific components but rather in terms of the end results of their application. The boundary is fuzzy and is evolving continually as new needs arise or take on increased emphasis. For example, the need for methodology management tools, mentioned above, has only become a priority in the past few years and is a result of both the complexities of today's design problems, the need for experts in relatively narrow disciplines to be able to use programs and techniques outside of their specialties without having to understand all of the details and, most significantly, the particular tool-based architecture that has evolved today as the architecture of choice for CAD systems. So at a particular time, there are many related facilities that can help in the implementation of a complex system and that are common to many different design styles. For example, tools for project management and documentation management, as well as tools to help the CAD developer, such as an effective software development environment, are important today and developments in these areas are described in Chapter 6.

3 DATA REPRESENTATION

3.1 Introduction

Before describing the various approaches that have been taken to the management of engineering data, it is necessary to introduce some common terminology and to describe some of the properties of engineering information that the data management facilities of a CAD Framework must be able to represent and manipulate. The Framework provides a set of mechanisms, or facilities, for modelling real-world information, and one of the most important issues in CAD Framework design is choosing a data model and corresponding implementation that is adequate for describing all of the information used by the design system, easily updated to new design styles and technologies, while remaining efficient and robust enough to meet the performance needs of engineering design.

3.2   Databases and Data Structures

Following Ullman [ULL82], we define a *database* as a collection of data "that is stored more-or-less permanently in a computer" and "software that allows one or many persons to use and/or modify this data" is referred to as a *database management system* (DBMS). Note that these are intentionally broad definitions. The concepts of database and data management system are treated very loosely in the literature and many different definitions of the particular features a system must support before it can be considered a database or data management system can be found. For example, a common characteristic that is used to distinguish a "database" from a "data structure" is whether the logical structure - the structure seen by a user or an application program - is different from the physical structure of the data - the particular arrangement of bytes and pointers used to implement the database on a permanent storage device. Of course, if the application program were not told that these two organizations happened to be the same, it could not tell that they were, and so the distinction, at this level, is purely syntactic.

Distinguishing between a database and a data-structure based on the way the data is represented in primary or secondary store is quite arbitrary. On the other hand, if a correspondence between the logical and physical descriptions of the data *is* required in the database, many other desirable attributes of the system would be difficult or impossible to provide, as described later, and so it might be considered a *poor design* for a database.

Other authors who attempt to define what is, and what is not, a database or a data management system often do so in terms of a particular model for a database architecture (e.g. the relational model) and corresponding data management system, excluding features that might be desirable if a different arrangement of data were used. Many of these distinctions arise because of the imprecise and evolving nature of the field in general and for these reasons, we choose the broad definition, qualifying particular database models, data management systems, and their implementations as "more useful" or "less useful," and pointing out features that are "desirable" or "undesirable" where appropriate.

Of course, one of the most important tasks of a DBMS is to provide an abstract representation of the physical data so that the user does not have to worry about where, or how, the computer system chooses to store the data.

3.3   Approaches to CAD Databases

The majority of successful database systems used in engineering applications to date have been *ad hoc* systems. That is, they are implemented to solve a particular aspect of the data management problem and cannot be adapted or expanded easily to deal with more general data management needs. More specifically, information about the database itself is built into the interface seen by the tools or by the end-user. For example, if the interface contained subroutine calls named `getNet,` `putNet,` `getGate,` and `putGate` for storing and retrieving netlist information, but did not contain facilities for dealing with other data, it would be very difficult to use the system to store mask layout information. On the other hand, if the interface contained the calls `getObject` and `putObject,` where the particular meaning associated with a given type of object was not known to the database per se, then the system could be used to manage many different sorts of data and could be adapted easily to deal with new data as well. The more general approach has some drawbacks, as presented later, but is essential in the engineering world, where the type of information to be managed and its relationships to other data is evolving continuously as we develop new technologies, as the need of the marketplace change, and as we learn more about the process of design itself. An important principle in the development of a successful CAD Framework is to assume the data model will change, probably quite often, during the lifetime of the system.

The use of a general-purpose query mechanism for dealing with the database simply defers the problem of associating meaning with the data - the particular types of data items and the relationships between data items that are supported in a particular database. In the example mentioned above, it must be specified *somewhere* that a particular object is, in fact, of type `net` and that it may have some objects of type `portInstance` associated with it. The particular types of data objects supported in a particular database and the relationships that may exist between objects is often referred to as the *conceptual scheme* or *schema* for the database. Many commercial database management systems provide a special high-level language, called a *data definition language* (DDL) for describing the conceptual scheme used in a particular database.

Another important feature of a database system is the query language, sometimes called a data manipulation language (DML), which allows the user (human being or CAD tool) to extract specific subsets of information from the database, via the DBMS. This is illustrated in Figure 3.1. The DML might be used to specify a request of the form *"highlight all of the logic gates in this design that have an output capacitance greater than 200fF"*. The

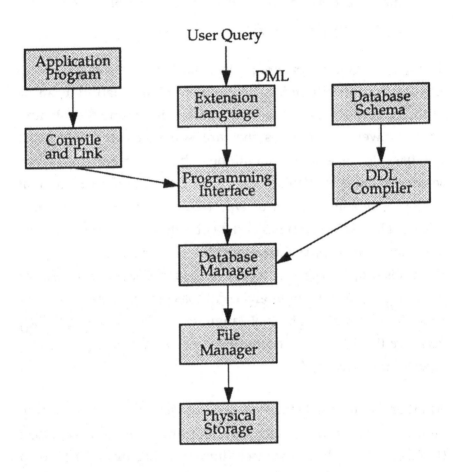

Figure 3.1: General Flow of a CAD Data Management System

fact that a database may contain an object called a *logic gate*, and the fact that it may have an attribute called *output capacitance* associated with it, which is able to store values like 200 femto-farads, is part of the conceptual scheme and was defined by the DDL. The expressive power of the DML (e.g. the number of DML statements needed to express a request), the complexity of the requests that can be made, and the execution efficiency of the request are important characteristics to consider when evaluating the effectiveness of a particular DML.

One significant difference between engineering data management and a common assumption for conventional data management is that in engineering applications many more queries are performed by application programs - the CAD tools - than directly by the end users. In a CAD Framework, the data manipulation language is often merged with the language interfaces to other parts of the system, such as user interface, design flow management, and history management, to form a common language interface to all of the facilities in the Framework. This language is then referred to as an *extension language*, as described in more detail in Chapter 8.

The data definition language uses a particular data model in which to express the conceptual scheme for a particular database. The data model consists of a mathematical notation that is used to express the data elements and their relationships, along with the set of operations that can be applied to the database to implement queries and other manipulations of the data. Of course, if a data model is general enough it can be used to represent any conceptual scheme. However, the choice of an appropriate data model in a CAD Framework has major implications, as described in Chapter 4.

3.4   Representational Issues

To illustrate some of the issues encountered in the representation of electronic design information, we will use elements of the circuit shown in Figure 3.2. This figure shows two logic **NAND** gates connected to form an **RS** latch. Rather than define what is meant by a **NAND** gate every time one is used, the gate is defined once in a master definition that is sometimes referred to as a *cell definition*. Each of the **NAND** gates in the figure is then referred to as an *instance*, or copy, of that master gate.

An important principle of engineering design is abstraction. That is, being able to represent a component or system by an abstract (less detailed) description. To be able to use an abstraction of an object, the user must first be able to "encapsulate" it. So a basic feature of any engineering data management system is the ability to encapsulate a collection of components and treat them as a single entity. In Figure 3.2, the **NAND** gates are abstract representations of transistors or some other lower-level and

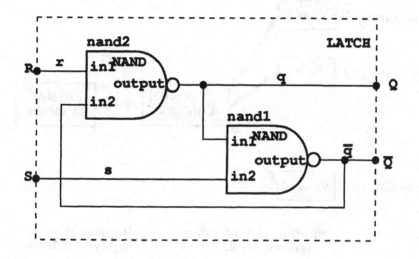

**Figure 3.2: Circuit fragment illustrating
re-use of component types**

more detailed description of the gate. They are then composed, with some connections, to form another component (the RS latch) which will be treated as a single, abstract component itself at some higher level of design. The general process of encapsulation and instantiation, applied in a nested way, results in a hierarchical description of the design as illustrated in Figure 3.3.

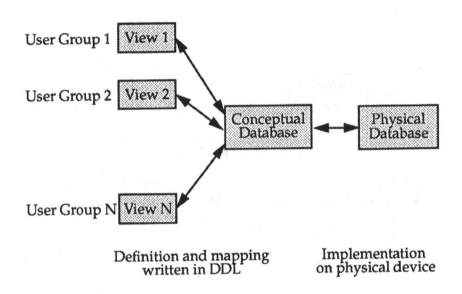

Definition and mapping   Implementation
      written in DDL   on physical device

Figure 3.3: Levels of abstraction in a database system

The instances of the **NAND** gate are connected together and to the input and output ports of the latch using nets. For example, the net called **q** is used to connect the output port of instance named **nand1** with port **in2** of instance **nand2** and with port **Q** of the latch. Ports on instances are copies of the ports defined on the master definition of the cell and so are referred to as port instances. Similarly, the ports defined on the master cell are referred to as formal ports, in line with their analogy to formal parameters in a procedure call. The ports **Q**, **Q̄**, **R** and **S** are the formal ports of the cell definition of **LATCH**. The name of an instance is used to distinguish port instances of the same name, e.g. **nand1/in1** and **nand2/in1** are different. Information particular to an instance or port instance must be associated with the instance directly (e.g. the output capacitance of a particular instance of a **NAND** gate), while information common to all instances of a cell is associated directly with the master of the cell (e.g. the ports that the cell defines). To complicate matters further, in the mask layout of the gate a particular logical connection point such as the output may be represented by multiple physical connection locations. For example, in standard cell designs connection point to signals are often available on both sides of the cell. Multiple connection points for the same logical port are called *port implementations* or *pins* and are illustrated on the right-hand side of Figure 3.4.

One of the most common mathematical tools used today for designing a conceptual scheme is the entity-relationship model [CHE76, CHE80] and we use it to introduce a number of engineering data representation concepts. In this model, the term *entity* is used in a very broad sense to mean "a thing that exists and is distinguishable." [ULL82][1] For example, a particular logic gate, net, rectangle, schematic diagram, or behavioral description would each be regarded as an entity. A group of similar entities forms an entity set. So all of the logic gates in a design might be regarded as an entity set. The adjective "similar" is used because the precise choice of attributes that are used to define a particular entity set is an important design decision to be made in representing the data in the database. Of course, entities may be members of multiple sets, for example a net may belong to the entity-set *nets*, as well as the entity-set *unrouted*. In modern programming environments there is a strong analogy between entities and instance variables and between entity set and classes.

Entities may have properties associated with them, called attributes, which associate a value from a particular domain of values for the attribute with the entity. For example, a logic gate might have an attribute *output capacitance* whose value, *200fF*, is selected from the domain "real-numbered values of type capacitance." Attributes can be used to distinguish entities that are alike. For example, the attribute **name** for a net in a circuit netlist could be assigned the unique value **output** to distinguish it from other nets.

---

1. The term "object" is often used in the same sense as "entity" although some users of the term "object" also imply an associated set of methods, or programs, which are used to manipulate the objects of a particular class.

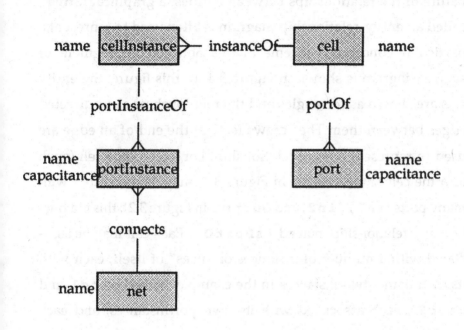

**Figure 3.4: Partial entity-relationship diagram for
netlist data representation**

A relationship among entities is defined as an ordered list of
entity sets and a particular entity set may appear more than once
in a relationship. The simplest form of relationship on two sets is
one-to-one, where each entity in either set can be associated with
at most one member of the other set. One-to-many and many-to-

many relationships can also be defined. Because of the complex nature of the relationships between entities, a graphical format, called an entity-relationship diagram is often used to represent a particular conceptual scheme in this notation. A fragment of such a diagram is shown in Figure 3.4. In this figure, the entity sets are shown as rectangles and the relationships as undirected edges between them The "crows feet" at the end of an edge are used to represent a *many* relationship. For example, a cell (such as a the cell named **NAND** in Figure 3.2) may be associated with many ports (**in1, in2**, and **output** in Figure 3.2); this is a one-to-many relationship, named **isPortOf**. Each cell may be associated with a number of instances or "uses" of itself, each with its own name (two instances in the example, named **nand1** and **nand2**), each associated with its own portInstances, and each portInstance associated with at most one net entity (for example, the portInstance named **output**, associated with cell **nand1**, is itself associated with the net **q**). This simplified fragment illustrates the use of the basic entity- relationship model.

Researchers have embellished this basic model in many ways, by being more specific about the *many* relationship, providing upper and lower bounds, and by assigning predicates to the edges of the graph, for example.

Other models have been used to describe engineering data, most notably the relational, hierarchical, and network models. Recently, so called "object oriented" approaches have been proposed, where the entity-relationship model is implemented most closely. The issues relating to the use of such models for engineering data are presented in Chapter 4.

3.5   The Nature of Engineering Design Information

Before a data model and associated implementation, complete with DDL and DML, can be evaluated for use in an engineering Framework, it is necessary to have a clear understanding of the nature of the information that must be represented and the requirements imposed by the users of the system, both tool and designer. A complete description of all of the issues and trade-offs that make engineering data management "different" from more conventional needs is beyond the scope of this book, but some of the most important characteristics are described in the remainder of this section, along with some of the more difficult issues facing the Framework designer today.

To begin with, the design of an engineering product often requires the cooperation of a broad range of specialists, each with their own needs regarding the particular data they wish to work with and with own their performance expectations. In addition, they do not wish to "see" the data associated with other aspects of the design. For example, in the design of an integrated circuit, the mask-level layout information consists of polygons on different mask layers while the gate-level schematic diagram or netlist consists of a collection of cells, such as nand gates or inverters, and a list of associated connections between the terminals or ports of the gates. The mask designer, concerned with relationships line the spacings between adjacent polygons on a mask, has no interest in the gate-level netlist but is very concerned that queries of the form "for each polygon on the mask layer named **POLYSILICON**, highlight all the polygons on that layer that are within one micron of it" - a two-dimensional geometric query involving tens of millions of entities. The logic designer, who has no interest in the mask layout representation of the design, is very concerned that all of the logic gates are connected correctly and might ask "for each logic gate, list any gates whose output ports are connected to more than five input ports" - a linear, connectivity-oriented request involving perhaps hundreds of thousands of entities.

In the database literature, an abstract model of a portion of the complete conceptual database is called a *sub-scheme*, or *view*. In fact, a view might contain information that is actually derived from the stored data on the fly. This concept is illustrated in Figure 3.5. In a general-purpose DBMS, a special sub-scheme DDL and DML might be provided. However, the implementa-

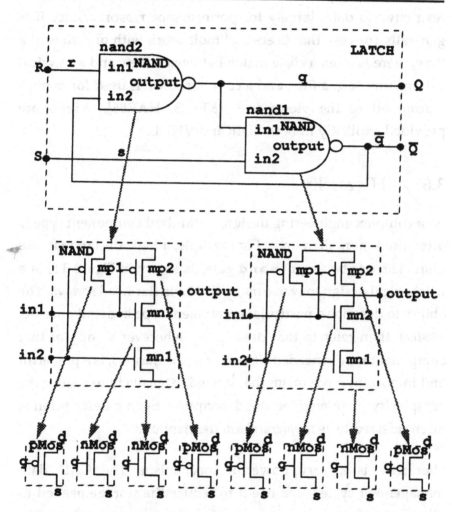

Figure 3.5: An example of an hierarchical circuit description

tion of views in practical engineering databases has been more restrictive to date, largely for performance reasons. Since it is generally the case that users and tools work with one view at a time, there is often a close match between a view and a physical storage unit (e.g. a file), and a common DML is used for manipulating all of the views (e.g. [KEL83, HAR86]). Views are provided explicitly in both EDIF and VHDL.

3.6   Hierarchy

In a complex engineering design, a standard component type is often used more than once. For example, a logic design may use more than one two-input **nand** gate, as shown in Figure 3.5, or a mechanical design may use many #18 1.5in flat-head screws. The ability to describe a particular component type in detail once (its *master*), then refer to that description wherever a copy of that component type is needed (*instances* of the master), is a powerful and important mechanism that is used extensively to reduce the complexity of an engineering description. Such a description is often referred to as an *hierarchical* description.

Hierarchy is a very powerful mechanism. The CAD data management system can use it to reduce the storage needed to represent complex designs, provided the same type of component is used more than once in the design. But it has its limitations and they can be a trap for the unwary. To illustrate a

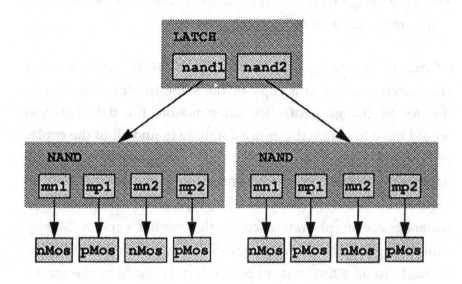

**Figure 3.6: An hierarchical design represented as an
instance hierarchy**

common problem encountered in the use of hierarchy, consider
the latch circuit illustrated in Figure 3.5. The latch contains two
instances of a cell called **NAND**, and since they are instances of the
same cell type, they contain the same sub-components. Each
NAND uses instances of **pMos** and of **nMos**. Each instance is
distinguished by its instance name, and the description could be
represented as an instance hierarchy, shown in Figure 3.6. Each

box in this figure represents an actual component in the design and if the design were stored explicitly in the database, each box would represent a stored entity.

Of course, having four separate copies of **NAND** seems a waste since each one is just a copy of the same master. Also, if one decides to change **NAND** for some reason, the data manager would have to search the entire database to find all of the copies of the gate to change them too. For these reasons, most CAD databases store the hierarchy more along the lines shown in Figure 3.7. Only one copy of each type of cell is stored, and instances are simply references to that master version. Significant savings in storage are achieved and any change that should be made to all **NAND** gates need simply be made to the master and they will be reflected immediately in all uses of the cell.

Now consider the situation where the designer has finished drawing the masks and wants to perform a simulation of the circuit using exact values for the parasitic capacitances derived from the layout. The value **Cout**, on the output of a particular **NAND** gate shown in Figure 3.5, must now be back annotated to the simulation view of the design. If the design were stored explicitly, as in Figure 3.6, it is clear that the attribute and its value would be associated with **nMos mn1** in **NAND** and **nand1** in the latch. But where should it be located in the storage scheme of Figure 3.7? If it is placed with the master on **nMos**, as shown

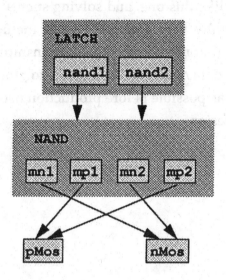

**Figure 3.7: Storage model in a database which uses hierarchy
to reduce storage requirements**

in the figure, then it is automatically associated with all four
nMos transistors, not what the designer wants. If it is associated
with the particular instance of **nMos mn1** in **NAND**, then it is associated with the instances of **nMos** in both of the **NAND**s in the
latch, again not what the designer wanted. The only way to store
this value correctly is to associate it with the full path name or
unique instance name for each component. In this case that
would be **latch/nand1/mn1/drain**.

Naive implementations of data managers for CAD systems often overlook issues like this one, and solving such a problem after-the-fact is generally very expensive, very inefficient, or both. Understanding the nature of the relationships that can be expressed in the data model and being sure to prototype as much of a data model as possible before production use are key issues in Framework design.

4 DATA MANAGEMENT

4.1 Introduction

While a storage management system manages a collection of data items and their relationships with no understanding of the actual meaning of the data, a design data management system uses knowledge of the structure of the data and its relationship to the design project to enforce constraints on the design process. For example, library management, design configuration management, and design consistency management are all tasks taken on by the design data management system.

Data management has been the most studied aspect of a CAD Framework to date and Katz [KAT85] presents an excellent review of the topic. Initially, data management consisted of basic file system protections on groups of files with occasional archival backup. Eventually it was recognized that the features offered by convention DBMS's might help in the management of design data. The results from this work are mixed, but the work led to the development of specialized databases for design data employing many of the features found in conventional database management systems.

The level of the data to be explicitly managed by the data management system can vary. Although some data management systems manage the individual data items or records at the level of a single transistor or net (fine-grain data), the majority of data management systems used in electronic CAD applications manage collections of individual data items without regard to the contents of the collection (coarse-grain data).

Many of the features of data management systems are described and the problems encountered in their use for CAD Frameworks are dealt with in this chapter.

4.2   Conventional Database Approaches

Many design databases have been built on top of commercial database management systems (DBMS's) [WON79, MIT80, ROB81, ZIN81, CHU83, JUL86]. However the use of conventional database approaches has not been very successful due to their poor performance or difficulty of use for engineering design [SID80, GUT84, KAT85]. Katz [KAT85] presents a number of the important features of conventional databases and compares them with the needs of engineering databases. They include efficient access to secondary storage, transaction processing, integrity maintenance, protection, concurrency, and crash recovery. In a DBMS retrieval of a single record of data from secondary storage is optimized. In an engineering problem large amounts of data need to be accessed quickly and so most engineering databases are managed in-memory today.

A key concept in the design of a data management system is that of a *transaction*. A transaction is a sequence of operations (such as the running of a CAD tool or a sequence of CAD tools) that, when complete, leaves the database in a consistent state. Transaction processing in a DBMS is based on the assumption that the transactions are atomic, quick, and modify a small amount of data. This assumption is generally correct for an airline reservation system or a banking funds transfer system, but is incorrect for the engineering problem, where the transactions can take hours or days (a design is checked out, modified, verified, and then committed back to the database). These *design transactions* [KAT85] are more closely related to the software world where a software module is checked out of a software revision control system, edited, compiled, tested, and then checked back into the control system. The engineering modification/verification (or software edit/compile) sequence can be iterated many times before the transaction is finally complete. While the amount of data in a standard DBMS transaction is quite small (an employee record), the amount of data processed in an engineering transactions is usually very large (for example the layout of a VLSI chip).

Conventional DBMSs provide pre-transaction and post-transaction consistency checks to verify that the database is in a consistent state. These checks are simple and static, such as ensuring the salary field of an employee record is positive. This contrasts with the consistency checks in the engineering domain which are complex and time consuming, such as ensuring the circuit performs to specifications. Such *transitional constraints* deal with the transition from one part of the design process to the next. In a DBMS, transactions are atomic and so there is no intermediate state preserved. In an engineering application where transactions can last hours or days, *checkpointing* must be performed in case of system crash. It is not sufficient to recover to the last consistent "saved" state of the database; the recovery must be to the last checkpoint. The design can use the versioning facilities to move back to the last "saved" state if needed.

In data management systems, protection against corruption or unauthorized modification to the data is very important. In a conventional DBMS, of primary concern are problems associated with concurrent access to the data. For example, in airline reservation systems, many travel agents may be trying to book a seat on the same flight at the same time. Interlocks must be in place to make sure that no more than one travel agent can modify the data. In engineering applications, the design is often partitioned in such a way that individuals can work on a single sub-component without strong dependence on the actions of other designers. Of course, the interfaces between the components must be defined and information regarding changes to the external constraints must be communicated, but the designer of an inverter, for example, is generally not concerned about the actions of another designer working on a **NAND** gate. While concurrency control is important in some engineering situations, access control - the ability to restrict access to design data for specific operations like reading, writing or modifying the data - is generally of more concern.

Perhaps the major factor limiting the use of conventional DBMS technology in the engineering world is performance. For example, Guttman [GUT84] used the INGRES DBMS [HEL75] to store geometrical IC mask layout data, and his experiments showed a slowdown from a factor of 3 in CPU time (a factor of 5 in elapsed time) over an in-memory CAD data structure for finding all geometries in a design, and a slow down of a factor of 20 in CPU time (a factor of 45 in elapsed time) for geometric queries, i.e. find all geometries in a particular subregion of the design. He found a number of problems and proposed solutions (some were specific to INGRES and some were generic problems).

Much of the performance problem can be attributed to an underlying data representation model that is not well suited to the type of data stored, nor the sort of queries engineering design tools make of the database. Many queries in CAD applications require transitive closure: for example finding all cells of a given type used in a design. VLSI designs using geometries that have two and three dimensions. Classical one-dimensional indexing schemes in a DBMS are not appropriate for the two and three dimensional data in VLSI designs. As a result, spatial queries are extremely inefficient, although some researchers have made progress in this area [GUT84, AST76]. The majority of efficient geometric search structures for CAD applications are in memory structures (quad-trees, k-d trees, corner stitching [OUS84]; see [ROS85] for a description and analysis of various in-memory region searching structures). Attempts at quick geometric search for secondary storage have been oriented towards points and not regions.

The relational model, which uses tables (called *relations*) of elements of a pre-defined format (called *tuples*) to represent data, is certainly efficient for representing large amounts of data that has a relatively fixed static structure (such as employee records or parts inventories), However CAD applications have large amounts of irregular data. If all **NAND** gates in a design had two inputs, a relation representing a two-input gate would be sufficient. But if gates can have any number of inputs, the model is no longer well-suited to the task. Record chaining and other ways around the issue must be used. Guttman [GUT84] recommended the addition of abstract data types to a relational database, better use of in-memory data, compiled queries, and tighter integration of the database system and the CAD application to improve performance. Other attempts have been made to extend relational DBMS to solve some of these problems, but the resulting DBMS is no longer truly relational and often moves into the *ad hoc* category.

Whereas Guttman used a relational database for storage of fine-grain data (for example individual geometric objects) and their relationships, Bennett [BEN82] used a relational database for the storage coarse-grain data and their relationships, leaving tools to interpret and access the fine grain data. Bennett used this approach successfully in an early version of the Mentor data manager.

Conventional DBMS's are optimized for accessing the current state of the database. The only history is an audit trail that is used primarily for verification and crash recovery. Previous states of the database are not saved. In engineering applications, it is critical that previous versions of data be saved. Having history allows designers to easily back out of bad design decisions, provides an audit trail showing how the design developed to its final form, and may be necessary for patent reasons (for example determining the first time a particular circuit architecture was entered).

In the past few years, attempts have been made to apply general-purpose object-oriented database systems to engineering design problems (see for example [BREU88, GUPT89]). Early work showed discouragingly slow performance, and ongoing work is focussed on replacing the generic routines provided by such systems with techniques specific to engineering databases [GUPT89]. Whereas an engineering database system today might be able to retrieve on the order of thousands of entities per second on a workstation, these general-purpose systems show performance in the tens to hundreds of objects per second. However, the object-oriented model overcomes many of the features of the relational model which restrict its efficiency and has the potential of being a better match to the engineering data management task. In recent times, the work of the object-oriented database companies has started to show considerable promise.

4.3   Physical Versus Logical Data Management

Many of the major performance problems in data management systems can be boiled down to a single issue - that of *name resolution*. Given the *logical name*, or *handle*, of an object, find the physical, stored data represented by the name. Designers and tools should deal with the data independent of where or how it is stored, based on some logical (domain dependent) organization. The logical name for a piece of data might contain some information about the way the data is organized (logically), like an hierarchical file name, or it might be an arbitrary but unique token that is translated to both a logical and a physical organization dynamically, when such information is needed by the user or the tool. Such tokens are referred to as *object identifiers*, or *OIDS* [WEIS86]. File names have the advantage of being quite understandable to humans. OID's have the advantages of generally being shorter than file names and independent of both the name of the object and a particular logical organization of the data. Whatever naming scheme is chosen, the logical references must be translated to the physical location of the data. Logical references can be context sensitive (full logical name of a reference depends upon the location of the parent, e.g. OCT [HARR86]) or must be unique across all designs, e.g. DOSS [WEIS86]. By using a synthetic name (OID) rather than a file name, the translation can be performed by the DBMS rather than by the operating system. OID's allow the easy migration of the

database since it has access to the translation facilities. In a file
name scheme, the translation is done by the operating system.
Translation of the logical reference to the physical reference can
be as simple as looking up the reference in a static table or going
through a multiple level mechanism where a dynamic table on
the local machine is queried, if the translation does not exist, ask
known servers on the network for the translation, and if that fails
broadcast the translation request to the network [WEI86]. How
the tables on the workstations and servers are invalidated and
updated is analogous to the cache consistency problem in multi-
processors. As data migrates from a server to a workstation
translations for that object in tables on workstations must be
invalidated and possibly updated. Also as servers containing
replicated data go up and down, the tables must be updated to
correspond to the current state of the network. Techniques
developed for multiprocessor memory management, such as
caching of name translations, the use of "snoopy" protocols
[GOO83] to reduce server contention, and replication of shared,
read-only data have been adapted for database name resolution.

Using logical references allows the physical location of the data
to change over time. A shared library may exist on multiple
servers on the network and if one server goes down the copy of
the library on another server should be referenced. When a
portion of a design is checked-out for modification, if should

move to the designers workstation. Heavily accessed designs should also migrate. All of this should be transparent to the designer.

Shared libraries and large designs rarely entirely reside on a single workstation and thus are distribution over a network of machines. In order to make sure that disruption of one or more machines does not stop work on a design, read-only libraries are usually replicated on many servers on the network. When a machine serving a currently referenced library goes down, the logical to physical translation facility modifies the translation table to point to another server. Note that this can also be accomplished on modern operating systems with file names by using symbolic links and remote file system mounts.[1]

In the early stages of a design, a designer may want to experiment with many different alternatives and be able to easily switch between them. By using a naming scheme that evaluates the location of the masters of instances at runtime the alternatives can easily be changed by changing run time parameters. This form of naming is called *dynamic binding*. The binding happens when a reference occurs rather than on creation (*static binding*). As the design progresses, a single alternative will be selected for use. At this point in time the references should be

---

1. For example, this is how shared libraries are handled on certain networks at Berkeley.

made static so they will not change due to runtime or environ-
mental changes. Although dynamic binding is particularly
useful early in a design project, where libraries and design organization change quite often, a dynamic approach can lead to
significant data management problems at the end of a complex
design. The dynamic nature of the name resolution mechanism
can lead to unpredicted side effects, where an apparently
isolated change to a specific reference can cause other references
to change at name resolution time.

Another name resolution and efficiency problem is caused by the
need for change propagation. As the design progresses, there
may be changes to a cell that causes inconsistencies in the database (such as changing the size of the cell or the number of
ports). In order to bring the portions of the design that use the
cells back into consistency, the changes must be propagated
[CHO88]. There are two different ways to propagate the changes:
immediate or *lazy*. In the immediate mode all references to the
changed object are immediately changed. If the entire database
is in memory and all references have been resolved, which is
rarely the case, then the mechanics of the immediate update
operation are trivial.

Normally, however, a few references will be in memory but most
will not, so the references must be located. If there are back
pointers from the cell to all instances this is straightforward. If

there are no back pointers then immediate update is out of the question. In this case lazy updating is used. In the lazy update scheme when a cell is referenced the information stored about it in the instance record (such as timestamps, size, number of ports) is compared with its current state. If there is a discrepancy, the instance record is updated. However, if the portion of the design with the discrepancy is not checked out for modification, this description will persist and the update will occur every time that portion of the design is processed.

Some systems [HAR86] take a middle road, in which all references that are currently in memory are fixed immediately and the rest are handled on the next reference to the portion of the design that has instances of the changed cells is processed.

Whether the changes are immediate or delayed only some can be automatically handled. In many instances the designer must be notified. For example, if a cell has a port deleted and there are connections to that port when that cell is instantiated some form of manual cleanup must be performed.

In order to reduce the amount of time it takes to do design, reusable components are designed for use in other designs. Examples are TTL parts in printed circuit board design, and standard-cell and macro-cell libraries in integrated circuit design. These libraries are different than other parts of the design. They are usually in a central location, replicated on other machines to handle network disruptions and bottlenecks, and marked read-only to protect against accidental modification. Since there may be many libraries that can be used in a design, it may be advantageous to name the references to the library cells so that library changes can be easily made. This type of naming is called *dynamic*. This makes it easy to change libraries by changing some library search path, but means that the entire design can be drastically and possibly fatally changed by a very simple change. Once the design has moved farther along the library that is used should not change, the references should change to static.

4.4   Managing the History of a Design

In engineering applications it is very important to maintain the history of a design [KAT86, CHO88]. This allows the viewing of the progression of a design, exploring alternatives, and the easy backout of bad design decisions. In engineering applications, history is normally kept as a sequence over time of a each individual design entity. Some systems just copy the entire design entity, some store try to just store the changes from version to version. These approaches are analogous to the versioning facility in provided in operating systems, such as Digital Equipment Corporation's VMS^{TM} operating system, which store copies of each version of a file and which have a mechanism for cleaning up old versions, and to a revision control system [TIC82] that keeps the current version and changes ("deltas") that must be made to recreate the previous version from the current version of the file, respectively. The latter approach is more difficult to implement efficiently for engineering design data, which does not conform to the simple linear record-oriented structure of a text file. In either case, the version management system must be able to support both linear versions of design objects as well as alternatives of any object, as illustrated in Figure 4.1.

Versions of single design entities usually have relationships to versions of other design entities and thus can not be thought of individually. The ability to keep collections of related versioned design entities together is another important requirement of a design data manager. These collections are called *configurations* [BEN82, KAT86], and they allow information about the state of an entire design to be recorded (called a *snapshot*). One major difference between a configuration and a version of a hierarchically organized design is that the configuration can manage data objects that are related to the design but may not actually be instantiated at the time, such as unused cells in a cell library, whereas the version only represents the data it uses.

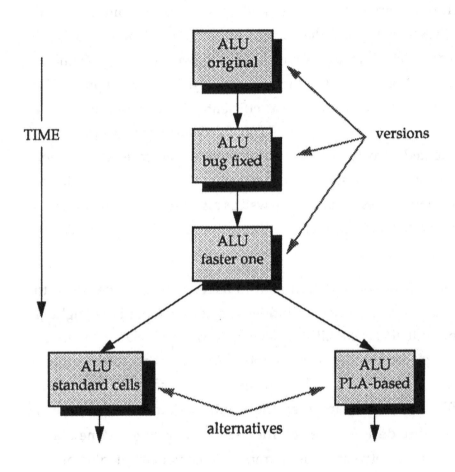

**Figure 4.1: The version and alternative approach
to history management**

The configuration-based approach to history management is illustrated in Figure 4.2. If a new configuration is created from an old one, it appears to the user as if all the design objects and their relationships have been copied to the new configuration. In reality, to save storage the configuration is implemented by storing the new versions of modified design objects, but only backward references to those components that did not change. Alternative design configurations can also be implemented by deriving multiple configurations from a single source.

4.5   Managing Multiple Users of the Data

Engineering designs are logically partitioned into sub-designs that are worked on separately with very little overlap. Since there is very little overlap, complex locking procedures found in conventional database systems are inappropriate and simple locking is sufficient. More important is access control. The system must be able to mark some designs read-only (shared libraries) and to control modification access to sub parts of a design. A design can have many designers, each working on different parts, but the entire design should not be modifiable by all the designers, each designer should only have modification permission for their particular piece of the design.

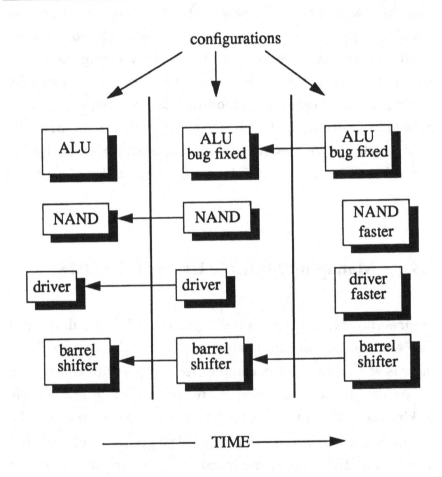

Figure 4.2: The workspace and configuration
approach to history management

Most design systems work on a check in/check out paradigm. In this technique, a designer checks out a portion of the design to a private area. Depending on the type of check out system, the portion of the design moved to the private area may be inaccessible for modification by others until checked in. On check in, various validation routines may be run to enforce consistency requirements.

4.6   Concurrency Control

When multiple users must access the design data concurrently, steps must be taken to protect the integrity of the system. For example, if two designers were able to modify the same gate layout at the same time and yet neither could see the changes that the other designer was making, chaos would surely be the result. The minimum protection that the design data manager must provide is the ability to restrict access to design objects on a selective basis. For example, certain objects may be marked read-only for all users except the cell librarian. Other objects might be tagged read-only by everyone except the owner of the object, who is permitted to modify it. Such simple access control mechanisms are necessary and even sufficient in many cases. Large engineering design tasks are often partitioned into sub-designs which are implemented separately and with very little overlap. As there is very little overlap, the complex locking

procedures found in conventional database systems are inappropriate in an engineering context and simple locking techniques
are generally sufficient. Another approach taken to access
control is to break up the libraries and designs into protected
domains known as workspaces [KAT86]. Workspaces can be
organized hierarchically: there may be a workspace for an entire
design while each of the major functional components of the
design may be contained in its own workspace. Workspaces
maintain information about who can check out the contents for
modification and they provide the necessary locking. According
to Katz [KAT86] a workspace should also be the granularity of
the design that is configured. Workspaces are often implemented
as a three-level hierarchy: *private* (can be modified only by a
single designer), *semi-private* (workspaces shared by a group of
designers on the same project), and *archival* (read-only workspaces shared by many different projects). A variety of
implementations of the workspace/configuration ideas have
been developed [CHO88, SIL89, WEI86].

There are cases, however, where multiple designers are
permitted to modify a single design object, provided the data
management system only permits one "writer" at a time. Most
design systems provide this concurrency control capability by
supporting a check-in/check-out paradigm, like that used in
source code control systems [TIC82]. The designer can check out
a portion of the design to a private area and then that portion

becomes "read-only" to all other users until the modified version is checked back in. References to the object from other parts of the design may be maintained with the old version or they may be updated in the new version of the object when it is checked back in.

Widya *et al* [WID88] present an excellent description of the problems associated with concurrency control in engineering database systems along with some solutions to the problems. The many forms of control other than those mentioned above include the use of hard locks, soft locks, lock transfer, and multiple writers. If a system uses hard locks, when a portion of a design is checked out, a lock is created (often an entry in the file system). When others try to check out the same portion of the design, the database detects the existence of the lock and the request fails. In a soft lock system, if a second designer tries to checkout some portion of the design that is already checked out, a warning is issued but the lock can be overridden. In a lock transfer system, the lock and the portion of the design checked out can be transferred from one designer to another, without checking in and then checking out the locked portion of the design. The most sophisticated of these schemes does not use locking explicitly, but rather uses the version-and-alternatives mechanism described earlier (Figure 4.1). In this approach, multiple designers can check out the same portion of the design for writing, but when they check the design object back in, each user creates an alternative of the original design rather than a sequential version.

Of course, merging alternatives into a single object that contains the union of the changes from the common source is not straightforward in general, but it is not really a Framework issue in any case. Simple locking seems to be sufficient for engineering databases, but Ecklund and Tonge [ECK88] have advocated the multi-writer technique for engineering databases. Other techniques involving the use of read locks to make sure that a modification is not made while the database is being read (a write can not occur until all read locks are removed) have also been investigated [WID88].

As the time between checkin and checkout may be long, and other designers may want to see the current status of the design, *nested* (or *tentative* [KAT85]) transactions can be used to obtain a snapshot of the design between versions.

4.7   Compatibility with Change

One of the major data management problems in the engineering world is that our understanding of the design process is continuously evolving, along with the technologies which are used to implement our products. New tools may be added to systems which still use an older version of the data model, and old tools may be needed to solve problems after the data model has evolved. These issues are usually classified under the heading of *upward compatibility*. The possible combinations of situations are illustrated in Figure 4.3. Changes to the data model can be classified as structural - those which affect the tool directly, such as changes in access methods or the meaning of a data value used by the tool - and non-structural, such as the addition of new properties to a design object.

Of course, by definition a new tool has no problem with new data and an old tool can read old data. In the case of a new tool which is trying to read old data, if the new tool does not depend on the changes to the data model there is no problem. If the new tool *does* depend on the changes, then the old data must be translated into a format that is suitable for the new tool. Missing data objects must be created and bound, values with new meanings must be translated, and so on. Many database systems provide facilities to support such updating of old data. The task can be performed dynamically and incrementally (only when the new tool is run), incrementally with update so that once the data has been converted to the new representation it is stored in that form for future use, or the entire database can be updated as a batch process. The second option is generally the most attractive for engineering applications.

	New Tool	Old Tool
New Data	*OK*	Translator or impossible
Old Data	Translator	*OK*

Structural Change

	New Tool	Old Tool
New Data	*OK*	*OK*
Old Data	*OK*	*OK*

Non-Structural Change

**Figure 4.3: The upward compatibility issue
between database and tools**

If an old tool is expected to read new data, a similar reformatting task must be performed if the new data involve structurally significant changes. If the changes in the new data model are orthogonal to the old model, the old tool should run with no problems. However the updating is implemented, it involves the equivalent of a translation or re-mapping step, and some systems provide support for developing such translators.

5   TOOL MANAGEMENT

5.1   Introduction

Tool Management involves the characterization and control of tools. Successfully running a CAD tool can be very complicated, as the designer needs to know many details, such as where the tool is located, the runtime environment required by the tool, how to invoke the tool, the command syntax of the tool, what translators need to be run beforehand, and what computer resources are required. At one stage the number of tools used in any particular design environment was small enough for designers to understand all of them. In this situation, with the absence of automated facilities for controlling the sequencing of tools, no characterization was necessary. The number of tools has grown considerably in the past few years, however, and it has become difficult for a designer to understand all of the tools available for design and analysis. Thus work has been done on building uniform interfaces for tools and providing consistent approaches to tool encapsulation.

5.2   Tool Characterization

One method of encapsulating tools is to build a generic engine for invoking tools and a language for describing the tools to the system. Among the features of a tool that can be characterized, the following are commonly required by tool management systems:

- tool name

- tool version

- physical location of the tool

- an icon for use by the tool manager user interface

- command line argument syntax

- help information

- computer resource requirements

- input requirements (which may require translation),

- output generated by the tool

- constraints to be satisfied before the tool can be run

- post-conditions to determine exit status of the tool

Tool managers are able to use this information to invoke and control the tool, removing the understanding of the details from the designer. Specification of the resources a tool consumes can be used to do load balancing or to find the machine best suited for the particular task. By specifying the types of the input and output files, the tool controller is able to invoke translators to convert to specified input files into the proper form and to convert the output of the tool into the requested output format. Constraint specification allows the tool controller to inform the designer (or design process manager) that there are constraints that have not been satisfied and must be satisfied before the tool can be run. Constraints are usually data consistency checks, such as making sure the data that was used to derive the input files has not changed since the input file was created, separate systems for this task have been developed [KAT86].

Various systems exist for characterizing and controlling tools, such as Cadence's Design Flow System which also encompasses methodology management, DEC's PowerFrame [BRO87], ULYSSES [BUS86, BUS89], CADWELD [DAN89] and MCC's MMS [ALL90, ALL91]. There are a number of systems that invoke and control tools, but ignore characterization facilities by requiring the tools to fit into a specific procedural interface, such as the RPC facility in the OCT/VEM environment [HAR86], the dynamic loading of the HAWK/SQUID system [KEL84], and the MAGIC system [OUS84], which requires linking the tool with the tool manager.

The CAD Framework Initiative has recently produced a Tool Abstraction Specification, which describes a textual language modelled after EDIF specifically for the purpose of characterizing CAD tools. The abstraction for a set of tools is provided as input to an abstraction compiler, to ensure that the description is syntactically correct. The compiler then stores this information into an internal data structure which may be directly processed by the Framework to provide access to the encapsulated tool. The abstraction format was demonstrated successfully at the ACM/IEEE Design Automation Conference in 1991.

5.3   Tool Control

Tools can generally be controlled by two methods: manual execution of the tool (including translation of input and output files, and satisfaction of constraints), or by tool managers. Tool managers are a special class of CAD tools that invoke and control other CAD tools. Tool managers can be integral parts of a design process system, as in ULYSSES, or stand-alone tools that can be used in a design process system, such as CADWELD. Tool managers can also invoke tools immediately or schedule them for processing (as in a batch queue manager). Dedicated tool managers, such as the PowerFrame system from DEC and CADWELD, have user interfaces specific to the tool management task. Some tool managers are actually the user interfaces to design systems, such as MAGIC, HAWK, and VEM, and they provide tight integration to the design system (user interface and database facilities) [KEL84], [HAR86].

Tool managers which have tool characterization facilities allow the control of any tool that can be described by the abstractions available. Tool managers without characterization rely on the tools conforming to some sort of interface (common input and output formats or procedural interfaces). This type of manager requires tools to be developed for the particular system the tool manager operates in or that 'wrappers' be created that convert between the standard interface and what the tool wants to see. In current systems the former is more useful, but as more integrated environments are built the former becomes more heavily used. Of course, this is an area where standards activities like the CFI can have a major impact on Framework development, by providing standard procedural approaches to tool integration and control.

A common set of assumptions for tool managers which do not use characterization is that tools read from standard input, write to standard output and standard error, and may be controlled at startup through a set of command-line options. These assumptions are particularly appropriate for Unix-based batch tools, though they may also be appropriate for some interactive tools.

5.4   Other Tool Management Functions

Tool managers can perform many functions beside basic tool invocation. These include:

- load balancing

- name serving

- consistency enforcement

- translation

- run logging

- status reporting

- license management.

Tools that do load balancing may work on a single machine or across a network. Load balancing on a single machine is usually handled by a batch queue. Batch queue software manages a sequence of jobs that are run in order, with the ordering based most commonly upon the priority and resource needs of each job. The designer (or design process manager) requests that a tool be invoked and the tool manager places the job in the batch queue. If the tool manager has access to machines on a network the tools (assuming the proper binaries exist and the tool has access to the input data) can be invoked on the "best machine," where best machine takes into account any special requirements of the tool (for example, runs well on a parallel machine, requires a lot computer time, uses lots of memory) and the load of the machine.

Name serving simply removes the need for the designer to know the location of a tool: it is a similar task to the name resolution issue for data presented earlier. The tool is referenced by a logical name - say "spice" - and the tool manager finds the location of the particular version of the tool which is appropriate to the task at hand, and initiates execution on the appropriate machine. Most tool managers allow the user to pick a tool from a list or menu of icons in a graphical user interface, and hide the details of the actual name and location of the tool binary.

Consistency enforcement is another important part of tool management. Before a tool is invoked the inputs to the tool are checked to verify that they are consistent and up to date, using some kind of constraint satisfaction mechanism. These constraints are usually data consistency checks, such as making sure that the data that were used to derive the input files have not changed since the input file was created, and which perform a similar function to that performed by software maintenance systems like the Unix **make** program [FEL79]. If the data are not up to date, the controller of the tool manager is informed. If the controller is the designer, then the designer can take the necessary actions to bring the data up to date. If the controller is an automated design methodology manager the tools necessary to bring the data up to date (if known) can be invoked. Sometimes consistency enforcement is handled by the data management layer that checks out or checks in design data before and after use [KAT86]. Some early tool control systems were based on adaptations of software maintenance systems for hardware design management (see for example [NEW81]), and recently separate systems for this task have been developed [TIC82].

Tools that have not been developed for an integrated environment usually represent input and output data as textual files. While some tools use common formats if they exist many tools use their own format. Therefore, in order to run many of the tools in a design system translators must be run in order to convert from the output format of the previous tool to the input format of the current tool. It is the job of the tool manager to determine (based on the types of the input files specified in the tool characterization) which translators need to be run and invoke them to produce the necessary input data. This process can be thought of as a simplified version of the problems handled by the consistency enforcement facility and can be folded into it.

Run logging is a useful facility where a closely coupled tool integration is not possible. The purpose of a run log is simply to document the activity of a tool, usually in a format which is both machine and human readable, in order that an audit trail of design activity is maintained. This can be used both to determine after the event which files have been read and written by a tool, and as the basis of some simple *post hoc* data management. The CAD Framework Initiative has defined a run logging format. Run logs can be generated by the tool manager because in general the tool manager will know which files are passed to a tool, and additionally some heuristic checks can usually find modified files after the run completes. The reliability of this process decreases as the interactivity of the encapsulated tool increases, however, and tools which have their own access to data stored anywhere on a network (such as editors) cannot be reliably encapsulated in this fashion.

Some tool managers are also able to provide status information on tool execution as a tool run proceeds, and thus allow finer grain control over the tool. Typical capabilities in this area include terminating execution of the tool, stopping and resuming the tool, and changing the execution priority of the tool.

Finally, a function which is of considerable importance in . commercial Frameworks is license management, which ensures that proprietary tools are only used in accordance with the contract between the user and supplier organizations. Licensing is usually handled through one or more license files, which contain encrypted information about tool access privileges, and one or more licensing *daemons*, which are small networked programs which continually maintain state information on license usage. A variety of licensing schemes are in use: the most common are *host-based* licensing, in which case permission is given to run a certain piece of software on a single machine, and *network* licensing, in which case permission is given to concurrently run a certain number of copies of an application on a network, usually defined by a list of host identifiers which are part of the encrypted license file.

In summary, tool managers provide a way of hiding the details of the actual invocation of the tool from the designer and also provide a consistent interface to the tools for an automated design flow manager.

6   DESIGN FLOW MANAGEMENT

6.1   Introduction

The notion underlying Design Flow Management (DFM), or Design Methodology Management, is that chip design is a process, involving a sequence of operations, performed on design data. DFM software attempts to capture and automate that process.

A DFM system may be viewed as a *meta-tool* in the CAD environment, both in the sense that it deals with other tools as data, manipulating them to meet some design goal which goes beyond the scope of any of the individual tools, and also in the sense that it "packages" groups of tools into higher-level entities which may be manipulated by the user as a single tool.

In the introduction to this paper it was stated that the criterion for success of a CAD Framework is that it reduce the time needed to develop or modify a CAD system such that the CAD system meets the needs of its end-users. In the case of design flow management, CAD system generation and modification is still part of the issue, but the tool has a function and significance which not only may make additional demands upon the CAD system developer, but which offers benefits to the end user which cannot be realized in the absence of a Framework. Design Flow Management may therefore be regarded in some sense as one of the fruits of a good CAD Framework.

6.2   Benefits

Design Flow Management offers two kinds of benefits: firstly through DFM it is possible to automate tedious sequences of tool invocations (for example an edit ~ netlist ~ simulate cycle); secondly it is possible to enforce development discipline within a design team - requiring management sign-off before committing library changes, or running DRC before approving layout changes. In addition to the above procedural benefits, it has been argued [SIE84] that explanation facilities, based upon the methodology and the state of the design are increasingly important as the tools become more autonomous. This need arises because under a DFM system, the user leaves some of the decision-making up to the Design Flow Manager, and then if things go wrong and user intervention is required, it must be possible to determine how the system arrived at its present predicament. In this context it is worth noticing that at least one worker has expressed concern [KAH87b] that with increasing "...automation and over dependence on CAD applications there is a danger that designers will fail to learn from the design experience." Explanatory facilities go some way towards alleviating this potential difficulty.

Design flow management has long been of interest in the ASIC world, where a large part of the business is logic replacement, and the typical designer is not highly skilled in the arts of chip design, as a means of protecting the user from methodological mistakes. The value of DFM is less obvious in the full custom environment, but it still has an important part to play in synchronizing the work of a team of designers, and automating multiple iterations and approval cycles.

6.3 DFM Dependencies

A DFM system depends upon most of the other services provided by the Framework. The most critical dependencies concern Tool Management and Data Management. As an analogy one might think of the interaction between the "make" utility, file system timestamps, and the compilation and linking tools (such as cc) which support software development. make enforces a sequence upon the execution of the tools, based upon file modification times.

The DFM software reads in or deduces a specification of a tool sequence (similar to the dependency graph generated by make) and it then activates appropriate tools, based both upon information about the state of the data (generated by the Data Manager) and upon the specified tool sequence, or program.

Some design management systems do not clearly distinguish between data management, tool management and design flow management. An example of a system which merges all three functions into a single tool is Sun's Networked Software Environment (NSE) [SWA88]. Although this tool was designed originally for the Computer-Aided Software Engineering (CASE) market, it has been applied to electronic design within Sun Microsystems.

The NSE allows different data objects, such as a schematic generated by a particular tool, to be defined, along with a set of appropriate methods which may be executed against objects of that type. This mechanism is used not only in conjunction with data access controls; it also allows the tool integrator to program methodologically derived checks to be applied to the tool and the data at runtime. An additional feature of the NSE which addresses tool and design flow management is the Link Services Database, which is a daemon-like facility which allows links to be established between data objects and arbitrary procedures. An example of the use of this facility would be to send mail to the users of a cell library whenever a change is made to the library. This very useful facility provides a general mechanism for instituting design checks, automatic tool executions, updates to design logs and the like.

6.4   Existing Approaches

A number of attempts have been made to provide Design Flow Management, either as part of an integrated CAD system, or as part of a stand-alone Framework. Notable in the former category are DEMETER [SIE84] and Ulysses [BUS87, BUS85]. In the latter category DEC's PowerFrame [BRO87] and Atherton Technology's "Software Backplane" [ATH89] are perhaps the best known.

In Digital Equipment Corporation's PowerFrame [BRO87], process-related knowledge is captured in the extension language, which provides, through a C-like [KER78] syntax, access to all the data management and user interface facilities of their Framework. Design flow management is therefore less an explicit provision of their system than a useful side-effect resulting from their system architecture. DEC calls the process of building these procedural cocoons *encapsulation,* and the term embodies some Tool management functionality as well as some design flow management.

Cadence's Design Flow System is another extension-language-based flow system: the design flow engine is driven by a set of data structures called *flowcharts* and *design steps,* which describe tasks and task dependencies using hierarchical directed graphs. Branching and looping capabilities add to the richness of the model. Each step (i.e. each node on the graph) is defined in terms of a set of procedures and data, defined in the SKILL [BAR90] extension language, which are activated by the design flow engine as required. The graphical model is supported by a graphical user interface which illustrates the flow graph and supports user interaction through direct manipulation.

The Microelectronics and Computer Technology Consortium's CAD program has developed a methodology management system known as MMS [ALL91], which uses MCC's extension language (Scheme) to describe tasks and processes in a declarative fashion. Some of the particular strengths of this system include the ability to distribute tasks across multiple hosts, and to gather together tool management and flow management activities. MMS takes tool and task descriptions and compiles them into an internal form which may be traversed by the MMS engine.

Paseman makes the point [ATH89] that what he calls work-flow control must "...respect and enforce organizational boundaries that are already in place" (p73). Many design managers do not use workstations, and their project records may not be on-line. Atherton Technology provide an interface to the outside world "...by allowing local policies to be implemented as message refinements and triggers".

SCHEMA [ZIP85], Ulysses, DEMETER, Sidesman [KAH87] REDESIGN [STI84] and VEXED [MIT85] take a knowledge-based approach to design: in each case the knowledge base is applied both in tool selection and for detailed design guidance. For a general discussion of a range of knowledge-based approaches to electronic design automation, consult [DAN87] or [BRE90].

These systems tend to work at a detailed level, applying small tools to small parts of the problem, and then gradually building up a complete solution. This is not compatible with current commercial CAD system architectures, which tend to be constructed as a relatively small set of "powerful" tools performing relatively independent functions. The designers of Sidesman stress [KAH87] "...the importance of the design of a complete environment for an 'intelligent' CAD system so that both rule- driven and conventional applications may be used to support designers."

VEXED divides the design flow problem into two parts: knowledge of implementation methods and control knowledge. The former describes legal operations, while the latter ascribes value to particular operations. In the DFM domain this maps (depending upon the granularity of the tools) to a distinction between tool activation knowledge and intentional knowledge about how best to proceed with the design. In addition, VEXED uses a DESIGN PLAN which records design decisions and their explanations, and also supports design replay for the exploration of alternatives. It seems likely that this kind of record, which is also maintained by the other knowledge based systems mentioned above, may well become an important part of future commercial design systems, both because of its explanatory value and its support for iterative design refinement approaches.

Rather than specifying tool relationships explicitly, either through procedures or rules, ELECTRIC [RUB87] schedules cooperating tools in a round robin arrangement, where the tools communicate through a change list and a common database. It appears that this should support tool interaction similar to that provided by a blackboard; however no explicit tool sequencing beyond the scheduler loop is used. This architecture thoroughly blurs the distinction between tool management, inter-tool communication, data management and design flow management.

6.5   Describing the Design Flow

There are a number of ways in which the intended design flow may be described to the DFM system. The nature of the description language has important implications for the flexibility of the DFM: in particular there are significant differences between a procedural and a declarative style.

The difficulty with a procedural style of data flow description is that it focuses on the "how", rather than upon the "what" or the "why". This means that each design flow, created by the tool integrator, specifies precisely the sequence of operations which constitutes a design flow, right down to prompting for user input, requesting tool execution, and checking result status. In terms of the software development analogy, this is like replacing makefiles with shell scripts. Not only is this kind of programming difficult, maintenance is extremely difficult, since interaction between tools and modules has to be described explicitly, making the addition or deletion of tools problematic.

A declarative style of design flow description has been explored by some workers. Ulysses describes CAD tool interdependencies in terms of preconditions, which are essentially assertions which must hold before a tool (or knowledge source) may be activated. It seems attractive to extend this notion to define each tool in terms of preconditions, actions and post-conditions: the actions are essentially procedural, while the post-conditions define the state of the design after the tool has run successfully. This allows a number of different execution models for the flow descriptions, as described below.

The designers of CHESHIRE [DEM87], an object-oriented integration system, take a rather different approach to design flow management: their coherence control is associated directly with the data objects, and is divided into three areas:

- **Data level** - access methods maintain data consistency

- **Application level** - a "local automaton" controls the evolution of a view within a particular tool

- **Inter-application level** - an automaton associated with each cell coordinates the evolution of the cell's views.

This style ties data consistency and flow control closely to the actual data, and shows once again how varied are the options open to the developer of design flow and data management tools.

VOV [CAS90] provides automatic creation of flow descriptions based on the notion of *design traces*. The idea is that as tools run, opening and closing files, they can leave a "trace" of their activity. This trace can be used to generate a graph of dependencies which can subsequently be used to provide records of design history as well as repeatable execution of combinations of tools. VOV automates not only collection of the information required for a flow, but also the re-running of pre-captured flows. VOV's traces are represented by a "...bipartite directed and acyclic graph, in which the nodes represent either design data or CAD transactions." [CAS90]. The design trace capture is implemented either by modifying tools to make calls to VOV, or by wrapping (i.e. encapsulating) the tools with scripts which keep track of file access at the beginning and end of each tool's execution.

Finally, a rule-based approach may be used to describing design flows, as we have seen with VEXED and others.

6.6   The Design Flow Engine

If the design flow is described procedurally, clearly the design flow engine, which executes the design to the flow specification, will be an interpreter for the design flow description language. However, if a declarative or rule-based approach is used, a number of alternatives appear. Of these, the most interesting are:

- **An inference engine,** which performs a search of the problem space guided by the rule base: without explanation facilities such systems may not be easy either to program or to understand in action

- **A simple procedural evaluator** (in the case of a declarative representation) which treats assertions as procedure calls

- **A functional evaluator,** which uses data management information to minimize the work required to achieve the current design goal. Such a system may be either eager or lazy, taking a conservative vs. an optimistic approach to previous design steps

- **A blackboard system,** where the state of the design and the goals of the user are modeled as assertions on the blackboard, and the design flow engine attempts to match the design state with the tool preconditions. Such an architecture actually supports all of the above models.

The important conclusion to draw from all of this is that the language used to describe design flows should not preclude the implementation of sophisticated programs for executing the flow. It is our view that procedural descriptions are significantly less flexible than declarative ones, because they say too much. Occam's razor is a valuable principle in CAD Framework design, because a design which specifies only that which is really required and understood does not eliminate appropriate extensions.

6.7   Standardization and Design Flow Management

Design Flow Management is very difficult to do without standards in a number of areas. Without a standard tool execution model the DFM system requires built-in knowledge about the execution environments of individual tools. Without a standard interface to data management it is difficult to determine the state of the design.

In the absence of a single standard database for electronic CAD, data interchange standards such as the Electronic Design Interchange Format (EDIF) [EIA87] have begun to make it possible to link groups of tools into sequences.

Another area in which emerging standards will make design flow management more valuable is user interfaces. The X Window System [SCH86] not only supports multiple simultaneous application displays, but it also allows display on a single screen of the output from programs running on multiple hosts. In such a heterogeneous, distributed computing environment, tools which perform "traffic management" among the tools make the designer's life somewhat easier. Finally, a single data model, shared by the CAD tools and the DFM in a single environment, will greatly simplify the tool management task, both at the tool integration level and the design flow management level.

To date, however, there has been little progress with respect to a standard model for describing design methodologies. CFI has focussed on tool management activities rather than attempting to define either a model or an interface to methodology description or management.

7 USER INTERFACES

7.1 Introduction

The user interface of a software system is that portion designed to interact with a human user. The focus of work in this field is in improving the communication between a user and the functional portion of a system. Good user interfaces gather and present information efficiently and effectively allowing a user to concentrate on the task at hand not on the software system itself. Development of such systems can be deceptively difficult. In this section, we highlight some of the problems in constructing user interfaces in the context of large Design Automation Frameworks and some solutions found in state of the art systems.

Research and development in user interface design is not limited to design automation. Since all application programs require human interaction to some degree, work in this area has been done in nearly all areas of computer research. Much of this work is informal and only manifests itself in the implementation of the resulting system. However, a specialized field of endeavor known as Human Factors has come of age to study this area exclusively [FOL84].

Conversely, development of user interfaces is heavily influenced by other unrelated areas of computer research. Advances in computing hardware have had profound effects on the design of human/machine interaction. Recent years have seen the development of new, more expressive input devices, faster processing speeds, and distributed computing via networks. These developments have given us the popular desktop workstation with multiple-window, pointer-based, graphic interfaces. Increasing complexity of software systems has also furthered effort in the user interface arena. Often, the control of such systems requires the user to assimilate huge amounts of data and make many complex decisions. Without an effective user interface, such systems aren't viable.

The major thrust of research and development of user interfaces is in two areas: improved application program interfaces and better interface development environments. Application programs are software a user invokes directly to accomplish a task. The user interface of such programs interact directly with the user and thus define the "look and feel" from the user's perspective of the entire system. Application programs are built on top of user interface development environments. Development environments allow user interface designers to experiment with new interfaces and build production interfaces quickly.

Recent development in application interfaces focuses on designing systems that can be used effectively by those with little knowledge of computers or programming and efficiently by those already familiar with the system. Such systems employ methods that allow entry and display of many different kinds of information including images, text, graphics and sound. These methods allow large amounts of data to be manipulated without overloading the user with too much information. Also, large software systems often consist of many independent components each with different user interface requirements. Modern systems try to provide common paradigms for interacting with all of these components in a uniform fashion. This minimizes the amount of low level system architecture information a user must learn (and more importantly, relearn) to use the system effectively. Furthermore, these common paradigms help to hide the often hard to use services provided by even lower levels of the software system (i.e. the operating system or even the underlying hardware). Eventually, such systems become tools that fade from the consciousness of a user in comparison to the work at hand.

Developments in the programming environments used to construct application interfaces are also an important area of research. Constructing a good user interface for any system is surprisingly difficult. User interfaces are judged subjectively by a body of users whose taste, knowledge, and experiences differ. Even the most careful designer can build interfaces that do not meet user expectations. Moreover, users often can't accurately describe what they need in an exact fashion. Even when exact specifications are available, the resulting system is often unsatisfactory. Implementing even simple interfaces involves a substantial amount of work both in design and implementation. Thus, modern user interface development systems must adapt quickly and allow (possibly radical) changes without massive redesign or re-implementation. Well designed architectures are the first key to such flexible systems. Recent work has yielded layered architectures that can be extended easily with little impact on other parts of the system. Embedded extension languages can also provide the necessary capability for quickly re-configuring a system. Finally, many systems offer means for a user to customize an interface directly and interactively. These techniques are presented in greater detail in section 7.3 on page 111.

In the remaining portion of this chapter, we present an overview of the state of user interfaces developed for design automation with an emphasis on electric circuits. Throughout much of its history, the area of electrical CAD has emphasized interaction with human users, many without detailed knowledge of software systems. A short history of these developments is given in the next section. Ideas used in developing interfaces to electrical CAD applications are applicable to other areas of CAD and general user interface development as well. Recently, the development of large numbers of design aids for the design engineer and integration of these tools through the use of Frameworks have made user interface design an even more important area of research for CAD professionals. The final part of this chapter will explore the state of the art in Framework user interfaces and possible future directions for such work.

7.2   History of Design Automation User Interface Systems

User interface design has played an important role in the history of electronic CAD research. New developments in this area have almost always been incorporated into leading CAD systems. Conversely, user interfaces (especially graphic interfaces) have been heavily influenced by continuing evolution in electronic design aids.

Three significant periods are apparent in the history of user interface design of electronic CAD tools. First, early batch oriented systems developed in the 1960s laid the foundation for later innovation in interactive interfaces. Second, starting in the early to mid 1970s, rapid advances in integrated circuit technology gave birth to the first interactive graphic interfaces used in electronic CAD. Finally, in the early 1980s, a boom in the development of automatic CAD tools for simulation and synthesis of electronic circuits led to development of Tool Frameworks with modern modular user interfaces.

Early electronic CAD tools were developed out of necessity. At the time, electronic circuits were constructed directly from specifications and diagrams drawn by hand. As designs became more complex, it became more difficult to check designs for correctness before the circuit was constructed. This problem became especially acute in the area of integrated circuit design. Early integrated circuits were laid out by hand by cutting shapes onto rubylith. With the invention of integrated circuit micro-processors in the late 1960s, the number of shapes in typical designs had grown to several thousand; well beyond the capability of humans to exhaustively examine for error. Thus, computer programs were developed to aid in the verification of these designs. Since integrated circuit design is a graphic process, the emphasis of such programs was graphic in nature. This emphasis would have a dramatic impact on graphic interfaces in the years to come.

In this period, user interfaces for these new computer aids were dictated by available computing resources. Most computer systems in this era were centralized batch-oriented facilities. Furthermore, most of these tools were developed in isolation with little influence from similar work done elsewhere. Thus, early user interfaces consisted of diverse card-image oriented data entry and line-printer oriented data output with no on-line user interaction. However, manipulation of graphic data using these kinds of interfaces proved too difficult and error prone. Off-line digitization tablets and plotting systems like those produced by David Mann were developed to remedy the problem. However, the batch nature of computing still ruled out direct graphic manipulation and interactive feedback.

Despite the batch emphasis of computer systems at this time, pioneering research in on-line interactive techniques became quite active [LIC62]. Cathode-ray tube (CRT) displays became available to researchers in the early 1960s. At this time, ground breaking work in interactive engineering graphics was done by Sutherland [SUT64]. His Sketchpad program is one of the earliest examples of on-line interactive manipulation of graphic images.

However, it would be another five years before systems developed for use in the integrated circuit industry would become available ushering in the next age in CAD user interface design.

Early experiments in interactive programs for manipulating graphics led to the development of dedicated graphics-entry workstations in the early 1970s. These commercial systems, developed by companies like CALMA, Applicon, and Computervision, allowed users to see designs graphically on CRT displays and directly manipulate them using digitization tablets or light pens. Soon after their introduction, these systems displaced the older text-based and off-line digitization schemes for preparing graphic information for later analysis and fabrication. Initially, these systems were very expensive, well over $130,000 per station. Thus, efficiency became the overriding influence on the user interface of these workstations. Unlike modern systems, quick learning time was not emphasized. On the contrary, dedicated technical personnel were trained specifically to quickly enter designs and were often kept busy around the clock to defray the large cost of the system.

Even though these systems emphasized input efficiency above all else, they contributed important technical innovations and improvements to the state of graphic user interfaces. Improvements to computing hardware played an important role in these developments. The greatest influence was the development of the mini-computer. For the first time, it became economically possible to dedicate a computer to servicing a small number of users interactively. Coupling these computers to raster displays allowed designs to be viewed in color and increased interaction with the user through direct manipulation of shapes on the screen with immediate feedback. Early versions of these systems used keyboard input for controlling all non-graphic aspects of the program. More accurate light pens and digitization devices caused user interface designers to concentrate on using the pointer more efficiently. Fixed on-screen menus and command annotations on digitization tablets required less typing and reduced hand an arm motion. Some systems even experimented with unusual forms of input. For example, some Applicon systems used sophisticated pattern recognition algorithms to recognize characters drawn freehand on the screen with the pointer. These characters were then interpreted as commands. Most of these features appear today in modern interfaces.

In the last five years, the advent of low-cost, high-resolution bitmap-based graphics workstations has dramatically changed the user interfaces provided to CAD tool users. In the early 1980s, inexpensive artwork entry systems [BIL83, OUS81, OUS84] and schematic entry systems based on well known inter- active graphics techniques [NEW73, FOL82] were developed that rivaled the capabilities of the dedicated graphics editors of the previous era. Instead of hiring and training dedicated

personnel to operate expensive layout entry systems, new
companies began providing these low-cost systems on the engineer's desk.

The state of computing hardware advanced rapidly in this
period. Lower cost personal computers and workstations made
it possible to provide greater interaction in CAD tool interfaces.
Very inexpensive raster displays with low-cost pointing devices
made it possible for CAD tool developers to write new interfaces
for applications traditionally done on text-based terminals or
through batch systems. The introduction of multiple-window
interfaces soon followed. These interfaces where based on the
pioneering work done in the late 1970s at Xerox [TES81, SMI82]
and made popular by Apple in its Macintosh [APP85] and Lisa
computers. These interfaces incorporated new features now
found in most modern systems: a window for each application,
pop-up or pull-down menu systems, forms-based input with
check boxes, toggle buttons, and text fields, and mouse based
manipulation of items on the screen. As designs became larger
and the demand for faster interactive interfaces for editing
graphics increased, new developments in data structures for
dealing with two-dimensional data came into use [BEN80,
ROS85, OUS84].

At the same time, the reduced cost of computing resources encouraged CAD tool developers to produce more automated design aids. Greater use of networks allowed the low-cost display workstations to be linked to the larger computers used for checking and processing designs. Entry and display of a wider class of information (not just graphics) became a necessary part of new CAD tool development. This led to current work in user interface frameworks.

7.3   Modern Framework User Interfaces

State of the art user interfaces developed as part of CAD Frameworks must meet a wide set of requirements. Input to such systems include high level problem specifications and parameters, design documentation and project management information, information for controlling a wide selection of automated design tools, as well as the graphics and design artwork handled by the older monolithic graphics editors. These new interfaces must also display new forms of output. These include intermediate design data (in both graphic and non-graphic forms), process statistics and management information, design documentation, status of CAD tools as processing proceeds, in addition to the final design data for fabrication and implementation. Furthermore, designers increasingly use applications outside the realm of the CAD Framework. Electronic mail, date-book systems, and document preparation tools are a few examples. Modern Framework interfaces must mesh with these other applications to provide a complete interface to meet all the computing needs of a designer.

The problems of designing comprehensive user interfaces that span many different applications is not unique to Electronic CAD Frameworks. These problems must be addressed by business and finance applications, engineering systems used in other disciplines, and general system support applications. Early work in this area was considered proprietary. However, recently a trend in user interface work has been to produce solutions that are released to the public domain. These solutions benefit all application developers by providing a common platform for developing compatible tools. These solutions are explored in the paragraphs that follow.

Current user interface architectures consist of several layers of software, each providing a higher level of user interface services. As shown in Figure 7.1, these layers consist of a graphics interface for controlling the display hardware, toolkits for building standard user interface components or widgets, a set of widgets for constructing the interface itself, and finally a high level programming interface used by Framework tools.

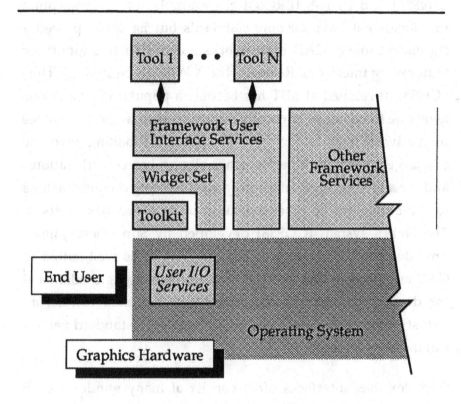

**Figure 7.1: Expanded view of user interface
components of a Framework**

At the lowest level, a graphics interface provides an abstraction
that hides the details of the underlying graphics input and
output devices. Early work in this area was done before the wide
acceptance of multi-window interfaces like those developed at
Xerox [SMI82] and Apple [MAC82]. The Graphics Kernel System

[ANS85] and PHIGS [BRO85] are examples of early graphics interfaces that have become standards but have not played a significant role in CAD Frameworks due to limited support for windowing interfaces. Recently, the X Window System [SCH86, SCH88] developed at MIT has become a popular choice at this level due to its heavy emphasis on windowing, wide acceptance by hardware manufacturers, its capability for making good use of a large class of high performance bitmap-based workstations, and for allowing application programs running on one machine to drive displays on other machines using a network protocol. The NeWS system [GOS86] developed by Sun Microsystems provides similar capabilities but may not be as well suited to CAD engineering due to differences in the imaging model used for displaying color. Work proceeds in this area toward an industry wide standard based on X but no such standard yet has full industry support.

Complex user interfaces often consist of many windows each displaying different kinds of information and each responding to different kinds of user input. The basic graphic interface provides the functionality to implement such an interface but at a level that requires an overwhelming amount of programming. Much of this programming is common to most applications. Toolkits, the next level in the architecture of modern user interfaces, have been developed to encapsulate the common portions of multi-window applications. These toolkits are based on the idea that complex interfaces can be built by combining standard interface components known as widgets. These components provide the labels, toggle switches, text input editors, menus, and other similar features found in multi-window interfaces. The underlying toolkit exports features for combining widgets on the screen, controlling their location, dispatching user input to the appropriate widget, and handling common resources like fonts and color.

Toolkits and widget sets are often considered a unit even though this need not be the case. For example, the standard toolkit for X [SCH88] specifies toolkit functionality without explicit reference to widgets. Many different widget sets have been built on top of it and some mixing and matching of widgets from differing widget sets is possible. Widget sets themselves define the look and operation of an application. Although a widget set is a collection of independent components, a cohesive set often defines conventions for uniform pointer and keyboard input, data display, and inter-component communication all using basic toolkit functionality. Two commercially developed widget sets, XUI from Digital Equipment Corporation and Open Look from AT&T, demonstrate this concept by providing components that when used together form an easily recognizable and consistent user interface.

Constructing applications using widget sets and toolkits can still be a complex task. Application designers must still specify the layout and interaction among widgets using somewhat cumbersome programming interfaces. Active research in general purpose user interfaces is now focusing tools for automatically building widget based applications from high-level descriptions. The pioneering work in this area was done in the middle 1980s by Apollo Computer Corporation in what is now known as their MOTIF system. Recently, Apollo has agreed to work on making this technology widely available possibly through a specification that may become industry standard. Current work focuses on specifying interfaces using a combination of a special purpose language description for specifying the interaction and control of widgets and interactive editors for layout and prototyping. Portions of this ideal have been constructed as part of the development of various widget sets and in proprietary systems. For example, Digital Equipment Corporation provides a language with its DECWindows widget set for combining the widgets into forms [DEC89]. However, there are very few widely available comprehensive application building tools in existence at the current time. Hypertext systems (such as the Hypercard system developed at Apple) allow the look of an interface to be constructed graphically and the semantics to be expressed in a simple programming language.

All modern CAD tool Frameworks use graphics interfaces and toolkits to integrate a wide range of computer design aids. However, systems in use today diverge in terms of their basic integration philosophy. The pioneering work done by EDA Systems [BRO87] and CADWELD [DAN89], uses tool encapsulation to allow older CAD tools with non-graphical interfaces to be used along with modern tools in a unified user environment. A modern windowing interface is then used to control the encapsulated tools. However, no attempt is made to fully integrate interactive tools or those with graphic interfaces. This may lead to inconsistent user interfaces. Comprehensive Frameworks with architectures like the one described in Figure 2.1 attempt to provide one uniform high level procedural interface to user interface services that can be used by all tools. Frameworks developed at Berkeley [KEL84, HAR86] and Cadence Design Systems use this approach. These systems allow tools become tightly integrated with a provided graphics editor. Tools can highlight graphics objects, obtain graphic and textual input, and carry out editing operations all within a uniform user interface. Cadence's Framework actually supports both techniques, in the interests of serving both the internal need for tight coupling between tools and their operating environment, and the external need to interface a wide variety of customer-supplied tools.

7.4   Future Directions for CAD User Interfaces

Major influences for future work in CAD Framework interfaces will come from increased standardization of toolkit and widget set functionality and by improvements to the programming interface provided by the CAD Framework standardization allows.

Standardization of toolkit and widget set functionality will allow Framework user interface developers to make greater use of these systems and thus provide much more powerful interfaces than those in existence today. Since these systems will be standardized, those developing other tools outside the realm of CAD Frameworks will also begin producing systems meeting these specifications. The result will be increasingly uniform computing environments where a designer uses CAD tools and non-CAD tools in an interchangeable fashion.

Future programming interfaces for CAD Frameworks will allow Frameworks to export a procedural user interface that is exported to all other CAD tools in the Framework. The key to the design of this interface is the realization that the user interface requirements of all tools can be met by providing the necessary facilities to construct interactive visual editors capable of editing any data stored in the Framework. Like advanced application building tools discussed in the previous section, this interface will provide a high-level abstraction above widget sets and tool-kits. Unlike general application building tools, this interface can be made simpler by tailoring the features toward those required by CAD tools. Widgets can be used to implement basic editing components like toggle switches, menus, and text editing fields (among others). Special purpose widgets for displaying design representation data would also be included (these widgets would provide functionality similar to that found in the older dedicated graphics editors). The programming interface to these widgets would provide higher level functions for creating and combining these components, means for calling user supplied functions when interesting operations in these editors occur, and high level functions for gathering input that can be called in user supplied functions. The data representation portion of the Framework provides policy routines for effectively manipu-lating design data. Together, the user and data representation interfaces would then be sufficient for creating a wide range of editors all tightly coupled to both the tool and the Framework.

Hypermedia is rapidly becoming a significant user interface technology, with obvious application to data browsing, docu-mentation and training tasks. Cadence's current Framework offering uses hypertext links to connect the Framework and tool user interfaces to on-line copies of the documentation, which may also be traversed directly using hypertext. One of the most important uses of hypermedia today is for accessing and understanding complex data organizations, and in the future we expect to see applications to simplify browsing and understanding the increasingly complex data hierarchies associated with large design projects.

These improvements will continue the general thrust of user interface development toward releasing users from the burden of understanding underlying system architectures and services. It is the translation of a user's ideas to and from a form that a computer program can understand that must be minimized. Early CAD systems required users to encode graphic information textually and interpret numeric output. Through the use of modern graphic interfaces, much less encoding is involved. Future systems will continue this trend until users no longer care about the underlying hardware and software architectures used to implement the system. Instead, users will focus on the problem at hand, not on the tool they use to solve the problem. This kind of tool interchangeability is a primary goal of the CAD Framework Initiative (Chapter 10).

8 EXTENSION LANGUAGES

8.1 Introduction

Since the earliest interactive software, it has been recognized that facilities for extending the capabilities of the system are valuable. Before programmatic means were available, Ivan Sutherland's SKETCHPAD system incorporated the notion of master and instance objects to allow the efficient repetition of a defined set of drawing operations [SUT64].

This kind of extensibility offers a dramatic reduction in the number of operations required for a drawing which involves repetition. When CAD tools first came into prominence in the electronics industry, the extremely high running costs encouraged the provision of macro capabilities simply in order that repetitive operations could be grouped and executed as a single command. The emphasis here was not on ease of use (the APPL-ICON and CALMA user was regarded as an expert technologist); rather the objective was speed, if necessary at the expense of having to learn a large number of commands performing subtly different functions. These macro facilities in general only allowed grouping of existing interactive commands: control structure and parameterization were not supported.

Some CAD systems provided command logs, or *flight recorders*, which were facilities for capturing the commands entered by the user in such a way that they could be rerun. The primary motivation for this capability was recovery from crashes or major errors: if design work is lost for some reason, it is possible to edit the command log manually to remove the offending commands, and then rerun from the previously saved configuration. In the Mentor environment, for example, users frequently use these log files as a mechanism for building the equivalent of keystroke macros, saving them in separate files as little "programs".

The keystroke macro approach has three major weaknesses:

- There is no control structure (branching, looping, subroutines etc.)

- There is no provision for parameterization of the macros

- There is no provision for local storage within the macro.

In addition, there is generally no on-line facility for documenting keystroke macros, so their value is generally restricted to a single user.

The designers of interactive software have recognized these problems, and gradually it has become the norm to provide a specialized extension language to allow users and tool integrators to customize the system in increasingly powerful ways.

To divert briefly from CAD to software engineering, it has long been recognized that extensibility is important to text editors, and Stallman's EMACS [STAL87] is a good example of the benefits of extensibility. EMACS is extensible both through keyboard macros and through a built-in Lisp interpreter. TECO, a popular character-oriented editor, is almost unusable without its macro facility.

8.2   Commercial Extension Languages in CAD

The success of facilities such as these have not gone unnoticed in the CAD world; commercial experience with extension languages includes the following examples:

- *AutoLisp* from AutoDesk

- *SKILL* from Cadence Design Systems

- *Genie* from Mentor Graphics

- *Ample* from Mentor Graphics
- *GPL* from Calma
- *E* from EDA Systems

AutoDesk's AutoCAD [AUT88], uses a dialect of Lisp called *AutoLisp*, derived from Betz' XLISP [FLA87]. XLISP is a subset of Common Lisp with object-oriented extensions. autoLISP makes no apology for its lisp heritage: extension language programmers work directly in the native lisp syntax.

Cadence Design Systems' Design Framework II uses a proprietary language called *SKILL* [BAR90, LAI86] which is loosely based on Franz Lisp [FOD83]. From a linguistic point of view, SKILL is a hybrid of Lisp semantics and C-like syntax. SKILL is very well regarded by its users, and it is generally believed that much of the power of the Framework and tools comes from the extensibility which SKILL provides. SKILL goes well beyond simply adding control structure, parameterization and variables: it also provides programmatic interfaces to the database, the user interface, and to tightly integrated tools. This means that it is possible to build substantially new functionality with the extension language.

Mentor Graphics' *Genie* is a proprietary language, based on Lisp, which is used primarily for procedural design. Despite being based on Lisp semantics, Genie supports two non-lisp-like syntaxes - one which is almost identical to C, and one which is more like a shell programming language. Genie was originally developed by Silicon compiler Systems, prior to their acquisition by Mentor Graphics

Ample is the extension language of Mentor Graphics' new Framework. Based loosely on C and Pascal (for compatibility with previous releases of Mentor's CAD tools), Ample provides a number of extensions which render it suitable for system customization, including a special command syntax which may be embedded in Ample programs, automatic memory management, and special data structures which are suitable for interfacing to the design database.

GPL [SMI75] was developed by Calma to provide extensibility for the GDSII product. The prevailing hardware constraints caused the entire language system to be shoe-horned into 16Kb; however the language was able to support an Algol-like syntax, polymorphic[1] functions and procedures, programming and command language capabilities, and both read and write interfaces to the database.

EDA Systems developed their language *E* as part of their Framework product [BRO87], to support tool encapsulation and user interface customization. E is approximately 80% conformant with C, and where possible adheres to the C semantics. Of the commercial extension languages, E has perhaps achieved the least success, and in current releases of the product, the use of E is de-emphasized. On reason for this is perhaps that C is designed as a statically compiled language offering support for low-level machine operations, while the requirements for an extension language are focussed around interactive use. E's poor performance, relative to compiled C was not offset by compensatory semantic or syntactic benefits.

8.3   Extension Languages Prototypes

In addition to the above examples which have been used commercially, there have been a number of other extension language developments of interest. Examples include:

- *OLAF,* from Honeywell Systems and Research Center

- *Common Lisp,* used by the Microelectronics and Computing Consortium (MCC)

---

1. A *polymorphic* function is one which is able to accept arguments of differing types and automatically handle the differences. This is provided in C++, for example, through a technique known as *overloading,* in which the programmer defines a function for each distinct set of allowable argument types, and the compiler uses static analysis to ensure the correct variant is called. Lisp system support for polymorphic functions comes through the ability to determine object types dynamically, and this is one reason for the popularity of lisp-based extension languages in the CAD industry.

- *Scheme*, used by Intel and MCC

- *LightLisp*, used by UC Berkeley's OCT/VEM toolset

As part of the Engineering Information System (EIS) project, Honeywell has developed an extension language called OLAF [KRU90] which uses lisp semantics, but an Ada-like syntax.

MCC originally used Common Lisp [STE90] as the primary implementation language for their CAD system. This proved too slow and difficult to support (especially given the concomitant requirement to use Symbolics Lisp machines, which were unfamiliar to the CAD clients of the system) and so in a second version of their system, MCC used C as the basic implementation language, while using Common Lisp to support system extensibility. This approach was much more successful; however the size of the Common Lisp system was felt to be too great in the end, and MCC therefore moved to Scheme.

Scheme [STE75, REE86] is a small, powerful dialect of Lisp developed at the Massachusetts Institute of Technology. Its primary advantage over Common Lisp is its small size, combined with semantics which support efficient implementation. Unlike Common Lisp, Scheme is not burdened by a large function library, and so it is suitable as an extension language engine, to be enhanced by CAD-specific function libraries. Intel have developed a prototype command language system based on Scheme. MCC are increasingly using Scheme in the role for which they previously used Common Lisp: as a mechanism for prototyping parts of the system where flexibility is more important than performance.

LightLisp is a subset of Common Lisp, developed by Wendell Baker to provide a means of prototyping applications for the Berkeley CAD tool environment. Versions of LightLisp support both database access (OctLisp) and editor customization (VemLisp). LightLisp typically functions as a server, providing new commands through remote procedure calls from the graphics editor, VEM. While LightLisp may be used to implement new commands, it is not a command language: this function is provided by the command processor within the VEM process.

8.4   Extension Languages Requirements

Given that one of our primary criteria relative to the goodness of a Framework is that it be possible to modify the system efficiently, it is clear that a good extension language plays a very important role. The extension language provides valuable insurance against changing requirements of the overall CAD environment.

CFI has performed a detailed analysis of extension language requirements [CFI91], and the following is a brief summary of the more significant ones:

- A safe environment for the programmer, in which errors are trapped gracefully

- Convenience is more important than execution efficiency

- Support for a range of programming paradigms

- Compatibility with the command language

- Run-time type checking

- Robust error recovery

- Automatic memory management

- Support for a good interactive programming model with built-in support for a good development environment

- A standard language as opposed to a proprietary one

- A simple syntax

- A good interface to programs and data implemented in other programming languages, especially C and C++

- Type extensibility

- Support for internationalization.

It should be noted that none of the candidate languages considered by CFI met all these requirements perfectly; however Scheme turned out to be significantly more acceptable than the other strongly supported candidate: C++ [STR87].

8.5   Design Issues for Extension Languages

Despite some notable extension language successes, it not easy to design a good extension language; nor is it easy to create efficient implementations.

Among the important issues are the following:

- Should a special purpose language be designed, or will an existing language suffice?

- Should the language be interpreted or compiled?

- How should debugging support be provided?

- Should the language directly manipulate database objects, or copies, or simply pointers to data?

- How does the language interface to the outside world?

- Should extension language programs be stored in the database, or should they be external entities, managed directly by the user through the file system?

The most successful languages to date are proprietary languages, generally based on Lisp semantics, which support an alternative syntax. This approach offers the interactive evaluation model and safe memory management which are characteristic of lisp, while providing a syntax, or "look and feel" which is more familiar to programmers with a background in the Unix world.

The advantages of an interpreted language are that it is easy to build a friendly development environment in which small pieces of code may be written and tested incrementally. The disadvantage is that interpreted programs run more slowly than their compiled equivalents, and in some applications where extensive database traversal is required, this significantly reduces the usability of the language.

8.6 General Applications of Extension Languages

Extension language applications cover the whole gamut of design activities, from control of the design environment to detailed manipulation of design data through extension language interfaces to CAD tool capabilities. Some systems do not distinguish between their command language and their extension language. In this case, the extension language supports macro and replay capabilities as well as interactive command entry. In most systems the extension language may be used to perform start-up configuration of the user interface and the tools.

User interface customization is an important aspect of modern CAD systems. Menus, dialogs, keystroke accelerators, default fonts and colors are all subject to user preferences, and these preferences are generally expressed through the extension language, perhaps in addition to direct manipulation methods of setting defaults. In addition, the extension language typically provides window management functions for the applications which are part of the Framework.

Even in the case where the command language and the extension language are separate, there is usually a mechanism to define commands in the extension language. AutoLisp, for example, which does not have a syntax which is directly appropriate for commands, allows commands to be defined as special functions, with additional arguments specifying the interactive behavior of the command.

8.7   Language-Based Design

So far, the extension language has been regarded principally as an extension at the user interface level, providing the end user with the ability to modify the behavior of the system for reasons

of end-user efficiency; safety or to provide new command functionality. However, an alternative view of the extension language
is that it allows the user to encapsulate design knowledge in
procedures, which when executed create new instances of design
components.

In the past, module generation languages have been separated
from general purpose extension languages; however it is not
clear that such a distinction is really required. A second distinction which obtains in current CAD systems is that between
Hardware Description Languages (HDLs) and module generation languages. However, now that Cadence Design Systems'
Verilog™ language, for example, is taken as input by the
Synopsys suite of synthesis tools, and is used to generate logic
designs, one might view the behavioral modeling language as
performing a function similar to that of a procedural design
language.

Clearly each application of a language within a design system
has specific requirements, not only in terms of evaluation model,
but also in terms of expressive requirements. This explains the
existence of a range of different languages within a single CAD
system. However, it would appear that if the right base language
could be found, one could build a variety of complementary
functional blocks by extending the language base. Some of the
commercial systems are implemented in this way.

8.8   Benefits of an Extension Language

The most important benefits of an effective extension language may be summarized as follows:

- **Openness:** An extension language can support tool integration by allowing the user to bind in new functionality through an extension language interface. Another important kind of openness which can be provided by an extension language comes through the provision of *callbacks*, or *triggers*, which may be called by the design system to perform some user-specified action at a particular time. Triggers are commonly used to customize user interfaces, and to allow user-specified actions when data is saved or modified.

- **Packaging:**[1] If all system functionality is made available through the extension language, it is possible to hide existing functionality as well as adding new functionality through control of the name space of the extension language, and by overriding default command and variable definitions.

---

1. The use of the term *package* for a module comes from Common Lisp; this technique of providing a controlled interface to a body of code and data is also known as *information hiding*.

- **Dynamic:** An extension language allows quick customization to meet specific needs - especially if the interface is interpretive. This is a most important benefit, as much of the customization which is typically performed in the CAD world takes place on a design in progress, in response to a particular situation which has arisen in the course of performing real design tasks.

- **Safety:** An extension language typically provides protected access to database and human interface, reducing the risk of either damaging the data structures or locking up the user interface

- **Encapsulation of Design Knowledge:** Procedures which either generate correct-by-construction design components or traverse designer-specific data structures increase designer efficiency.

8.9   Future Directions

In the future, we may expect to see more sophisticated uses of user programming features in CAD systems. For example, MCC's C Module Editor (CME) uses graphical programming to specify both constraints and iterative directives. The WireLisp [EBE89] system allows the designer to freely combine textual (programmatic) and graphical styles of design description.

Another method of programmatic system extension involves what the developers of LOOPS [STE83] call *active values*: attaching code to data, such that when the data is accessed, the code is run. This allows user constraints to be installed very simply, and goes some way to creating an "intelligent" database, where behavior can be stored with the data. Mentor Graphics' recently announced *Decision Support System*™ is based on a spreadsheet model, though it provides some computational capabilities which are traditionally associated with extension languages.

In fact the notion of storing code with the data is one which has gained popularity through the increasing visibility of object-oriented techniques. In an object-oriented system, when a message is sent to an object, it is unimportant whether or not the message is implemented by code or by pre-calculated data. This blurring of the distinction between code and data is familiar to the artificial intelligence community, where "late binding" of behavior to symbols is regarded as a valuable technique.

Finally, it is interesting to note that the CAD Framework Initiative's Architecture Technical Subcommittee has recently selected Scheme, a language derived from Lisp, as the basis of its standard extension language. This will provide a consistent environment for system customization from one vendor's Framework to another. Scheme will be augmented by an alternate syntax, which will make the language more cosmetically attractive to users who are not familiar with lisp. In this respect, the CFI solution closely resembles SKILL and Genie.

9 IMPLEMENTING A CAD FRAMEWORK

9.1 Introduction

There have not been many success stories to date in the design and implementation of CAD databases, let alone CAD Frameworks, although hundreds of millions of dollars have been spent trying to achieve this goal. We believe that there are important reasons why this is so and that the reasons have very little to do with what the designers are trying to build and are almost entirely concerned with the approach they take to the design and implementation of the system.

The design and implementation of a CAD Framework is a very complex task for a number of reasons. Firstly, a framework consists both of interfaces and implementations, the specifications of which are interdependent. Secondly, there is no "theory" of framework design upon which to base specifications. Instead there is a great deal of informal, empirical information about the requirements for the various framework components. Framework design can therefore be viewed as an optimization problem where the objective function is extremely difficult to calculate, and where it is additionally extremely difficult to determine the sensitivity of the objective function to individual system variables, and where the variables interact quite strongly.

A final issue which militates against complete framework specification before implementation has to do with the rapid pace at which the technology of electronic product design is changing. Not only are new tools being continually developed, but the hardware, the distributed environment and the operating system are all undergoing continual refinement. In addition the end user's requirements are changing as new kinds of tool become available, with associated representational requirements.

We believe therefore, that it must be a principal assumption of the framework developer that today's best solution will not meet tomorrow's need. Rather than abandon the venture as being too difficult, however, one has simply to ensure that the framework architecture supports extension and is highly modular, in order to allow gradual replacement of components as they become obsolete.

These difficulties in specification are not new to software engineering. There are a number of domains in which it has been long recognized that specifications in the abstract are unlikely to result in successful implementations. The rapid prototyping environments beloved of Lisp and Smalltalk programmers, for example, exist for precisely this reason. Two more interesting examples are the processes adopted in the development of MIT's X Window System [SCH86], and the Common Lisp standard [STE90]. In each of the latter cases, not only was development incremental, working through a number of prototype implementations and releases, but it was also highly distributed, involving principally electronic communication among a scattered community of experts. Both of these efforts have led to the creation of industry standards; however neither software system is without its critics. This is the nature of a democratic process.

Of late two efforts have been initiated to develop CAD Frameworks through this kind of successive refinement process. Firstly the Microelectronics and Computer Technology Consortium (MCC), has adopted a similar methodology to that used by the X Window System and Common Lisp groups, dignified by the acronym CODEM, which stands for COoperative DEvelopment

Method [BAR88]. Secondly, the CAD Framework Initiative (CFI) is attempting to specify CAD framework standards through a similar cooperative process. The standardization process being undertaken by CFI is described in more detail in the next chapter.

9.2   The CODEM Approach

The traditional "waterfall" model for software development involves a cycle of specifications and reviews between the developer and client. This approach makes sense under the following conditions:

- The client knows precisely what is required

- The product is not needed for some time

- The product can not be acquired by any other means

- The client is a unified entity

- The requirements will not change significantly during the life of the project.

- Unfortunately, these conditions do not fit well with CAD Framework development.

The conditions under which *CODEM* succeeds are quite different:

- There are many potential clients with similar, though not necessarily identical needs

- Some software exists which solves part of the problem and which can be used as a common starting point

- An organization exists which can serve as a focus, both taking responsibility for managing communication between the co-operating parties and for integrating and distributing the emerging software system

- Networked computers are available to all participants, supporting bulletin-board real-time message handling between the participants

- The software selected as the starting point is modular, with well defined inter-module interfaces.

- The selected software is available to all interested parties.

These are the conditions under which CAD Frameworks are beginning to emerge, both through explicit application of the methodology among MCC and its shareholders, and also as the CFI meets and communicates to develop framework standards.

The *CODEM* approach replaces the typical "waterfall" model for software development with a loop involving three steps:

1) Build a working prototype
2) Determine the most significant weakness with the prototype
3) Develop a solution and return to Step 1.

This process continues until the effort to improve the system outweighs the benefits. In the MCC case, the Berkeley Framework [HAR86] was chosen as the common starting point, because of its modular architecture and use of standards (e.g. Unix, the X Window System, Remote Procedure Call library).

There is a difference between the approaches taken by the two groups, however: the MCC group are convinced of the importance of a working prototype as a check on the viability of design decisions (they chose OCT/VEM/RPC from UCB), while CFI is focusing on interface standards, and regards implementations as secondary.

In retrospect, the *CODEM* approach was not as effective at MCC as had been hoped. The principal reason was probably the lack of committed resources from the MCC shareholders, which meant that the benefits of real-time dialogue and widespread use of the developing technologies were not realized.

9.3   Commercial Frameworks

Cadence Design Systems, Mentor Graphics and Viewlogic have all been going through the process of releasing framework products over the last two years. In each case, the companies have experienced difficulty in releasing products with sufficient performance, functionality and overall performance to satisfy the customer. EDA Systems was purchased by Digital Equipment Corporation after struggling with a second version of its PowerFrame product. Interact's framework product was sufficiently poorly received that the framework development team was disbanded and the product discontinued.

Part of the difficulty is that the industry as a whole is experiencing what Brooks [BRO75] refers to as the "second system syndrome". This is a situation in which there is some experience with the requirements for a product, and the developers become excessively ambitious in the specification of a successor. This leads to solutions which are too complex, too large and unwieldy, and finally too difficult to maintain.

Despite the obvious difficulties experienced by framework developers, user enthusiasm for frameworks is at an all-time high, as evidenced by the strong support for CFI.

10 THE CAD FRAMEWORK INITIATIVE

10.1 Introduction

No description of CAD Frameworks would be complete without a discussion of the standardization work being promoted by the CAD Framework Initiative (CFI). This grassroots organization has not only demonstrated beyond doubt the deep belief in standards which is shared by CAD users and vendors in both the Systems and IC markets, but it also has demonstrated remarkable progress in the first three years of its life: both in the production of standards specifications, and in the creation of live demonstrations of interoperability between frameworks and tools at the Design Automation Conferences of 1990 and 1991. Increasingly strong financial support augers well for the continued success of CFI.

Those who have worked closely with CFI recognize both the extraordinary level of progress which has been achieved in three years, and also the frustrations which come from the negotiation process. In this chapter we examine the origins of CFI; its participation; the organizational structure of the organization, and the technical activities undertaken by CFI.

10.2   The Origins of CFI

The CAD Framework Initiative was the brainchild of Motorola and EDA Systems, Inc.[1], the first and possibly last company to build an entire business on the framework concept. Such an organization was attractive to EDA Systems also because of the publicity and interest which was expected to surround such an organization. Motorola, an early purchaser of the EDA Systems product - proposed the standards body on the basis of their experience of the high cost of tool integration. These two companies sponsored the inaugural meeting.

CFI was inaugurated by a meeting held in Santa Clara on May 23rd 1988. At this meeting, thirty-eight companies were represented. Several companies spoke about the difficulties they had experienced in building and managing design environments.

---

1. EDA Systems Inc. has subsequently been purchased by Digital Equipment Corporation

These motivational speeches struck a chord with the audience, and it became clear that the issue was well understood - even if the solutions were not.

Among the early supporters of CFI, the Microelectronics and Computer Technology Consortium (MCC) was one of the most influential. The MCC CAD Program had at that time been in existence for some years, and had produced a prototype software system of great complexity embodying a sophisticated object-oriented architecture for data and tool management, along with a user interface and integrated editing tools. This technology having failed to meet the real needs of their shareholders, MCC had embarked upon a new framework development plan, building on software from the University of California at Berkeley: the OCT/VEM system [HAR86]. MCC was strongly interested in making a visible contribution to the CAD and semi-conductor industries, and they rapidly took on leadership roles in the fledgling CFI organization.

As it has matured, CFI has continued to receive strong support from CAD vendors such as Cadence Design Systems, Mentor Graphics, Racal-Redac and ViewLogic. These companies are motivated both by the need to meet customer expectations with regard to standards, and also by the perceived benefits of tool interoperability for both internal development and external linkage.

Hardware vendors - especially the major players in the engineering workstation market - have also taken a strong interest in CFI. Digital Equipment Corporation, Hewlett-Packard, IBM and Sun Microsystems, for example, are active both because of their interest in providing hardware and software products which optimally support computer-aided engineering and because they are chip and system designers, standing to benefit from the standardization efforts. As has been suggested previously, some of the services we associate with a CAD Framework could well be provided as part of an engineering computer's operating system.

A third class of participants is the ASIC system provider. Typically, an ASIC development system involves a well-defined design methodology, related to a specific set of manufacturing technologies, supported by a carefully integrated set of CAD tools. The ASIC designer is frequently less expert in CAD tool use than a professional integrated circuit designer, and such people need a design environment which reduces the opportunity for error. The cooperative nature of ASIC design, where the ASIC vendor and the customer together produce the final product, also requires clear direction for each contributor to minimize the risk of errors associated with miscommunication. Building such an environment from a heterogeneous set of tools - some purchased, some specially built - remains an expensive, tedious and error-prone task. Subsequent modification of the environment can involve changing thousands of lines of integration code. ASIC system providers are keenly interested in standards which will simplify tool integration, data management and inter-tool communication.

Finally, end users have become increasingly concerned with CFI. At the June 1991 meeting, held in conjunction with the Design Automation Conference in San Francisco, a CFI Users Group met for a full day, to hear about CFI's progress, and to offer its special perspective to the organization at large. This is an important development, because it provides a mechanism through which CFI is kept honest with respect to the needs of the final customers of framework technology.

Thus, over three years, CFI has grown to take in hardware and software providers; system integrators and end users. It is an extremely democratic organization which is making a strong impact on the entire electronic CAD industry.

10.3   Goals and Deliverables

CFI's goals are centered around achieving tool interoperability. This will be accomplished by defining a set of programming interfaces with which tools may be integrated. Each part of the framework architecture is defined both in terms of its function and its interfaces, in order that tools may reliably be constructed to use the interfaces and thus be integrated with framework services.

CFI does not attempt to specify the detailed architecture of a framework, because it is believed that this is an area for research and the basis for some competitive differences between commercial products. So long as a framework provides the functional capabilities, represented by the standard set of interfaces, it may be regarded as being CFI-compliant.[1]

The first sets of formal deliverables, known respectively as *CFI 1.0* and *CFI 2.0*, are scheduled for release towards the end of 1992 and 1993 respectively.

10.4   The CFI Organization

The structure of CFI was developed at the inaugural meeting, and until mid-1991 went mostly unchanged. It is detailed in Figure 10.1.

---

1. At this writing, the notion of *CFI compliance* is not well-defined, because there are no fully ratified standards. The typical practice of companies who wish to underline their alignment with CFI is to declare their conformance with a draft standard, especially those used at the DAC demonstrations.

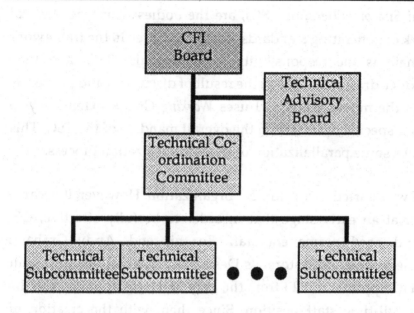

Figure 10.1: CFI Organization

The *CFI Board* is the formal steering group within CFI. It is populated by nine executives from the companies which make up CFI's membership. The *Technical Advisory Board* is made up of representatives from the academic world, who are consulted periodically on questions of approach. The *Technical Coordination Committee*, or *TCC*, is made up of the chairpersons of the Technical Subcommittees. The TCC acts as a gate through which draft standards must pass before going to vote by CFI as a whole; it also performs a tactical and organizational role in facilitating progress within the Technical Subcommittees. Finally the *Technical Subcommittees*, or *TSCs*, are the bodies which do the real work of generating standards. Each major area in the framework domain is the responsibility of a particular TSC, and they produce draft standards as the result of discussion and proposals from the membership. CFI uses *Working Groups* extensively to tackle specific tasks within the overall mandate of the TSC. This allows some parallelization of the standardization process.

CFI was started as a volunteer organization. However, it became clear at an early stage that in order to be fully effective, CFI would need permanent staff. To this end, Andy Graham, formerly head of Motorola's Design Automation Business Unit and a supporter of CFI from the very first, was appointed to the first full-time staff position. Since then, with the creation of corporate sponsorships, further appointments have been made. The permanent staff perform a number of important functions: marketing CFI; planning and coordinating progress and deliverables; providing technical and logistical support to the annual DAC demonstration projects.

For the first eighteen months, CFI meetings were held quarterly. This proved to be insufficient, however, and so a program of eight technical meetings each year with separate meetings for the TCC and the Board was developed. Meetings typically last five days, including TSC, Working Group and TCC meeting time.

During 1991, the notion of *Pilot Projects* was developed as a means of obtaining practical verification of the utility of the standards proposed by CFI. Each pilot project involves CAD tool developers *and* users, and are intended to progress through specification, prototyping and demonstration phases. This emphasis on practical work in addition to the meetings which characterized CFI's early work is a welcome progression, as it allows for debugging of proposed standards by both developers and users in the context of a real need.

The need for CAD Framework standards is by no means limited to the United States of America. European and Japanese members now represent 19% and 25% respectively of the total membership. While in general communication between the US and other groups has been less than ideal, an annual European meeting to which the US membership is specifically invited has improved things somewhat. Framework development in Europe is primarily taking place within research organizations - particularly JESSI - while there is strong commercial representation within the US.

10.5   Tangible and Intangible Benefits

Among the tangible benefits, the draft standards and the demonstrations clearly show the value of CFI's work. These are the achievements in which one most easily sees the importance of

CFI. There are also a number of intangible benefits which are worth noting. CFI causes the major players in the CAD framework business to come together, several times each year, to discuss the future of their discipline. This has led to a great deal of communication between groups who for competitive reasons would have spent little time together without the impetus of CFI. From this has gradual emerged a common vision of frameworks, shared not only by suppliers of the technology, but also by the user community. This is critical to the creation of a mature framework business, as well as helping to ensure that the needs of customers who wish to use tools from more than one supplier are better met.

10.6   Technical Activities

The material included in this section is necessarily based on the current status of CFI, and the status of each technical group will change over time. The purpose of including this information here is to provide a picture of the kinds of things which the technical community regards as important. The most significant difference between the work described here and that covered by the remainder of the book is that this work is rooted in a desire to find solutions to framework problems which are viable as commercial standards, rather than vehicles for research. This is not to suggest one is more important than the other; but simply to emphasize that the goals and therefore the methods applied are somewhat different.

Architecture

The original set of TSCs did not involve a group with responsibility for the overall architecture of the Framework. This lack soon became evident, however, as it was realized that the interaction between the services provided by each component was of great significance. To this end a number of documents describing the overall Framework architecture at a conceptual level have been developed. Over the life of CFI the detailed picture of the relationship between framework components has changed; however CFI's reference architecture diagram (Figure 10.2) has much in common with Figure 2.1.

Design Representation

The Design Representation TSC is concerned with the structure and semantics of electronic design data. In many respects this TSC has been the most successful. The primary reasons for this are twofold: firstly it is very clear to everyone that a common schema is a powerful aid to interoperability, and secondly this is an area which is relatively well understood. Every CAD vendor and most sophisticated users have implemented design databases at one time or another.

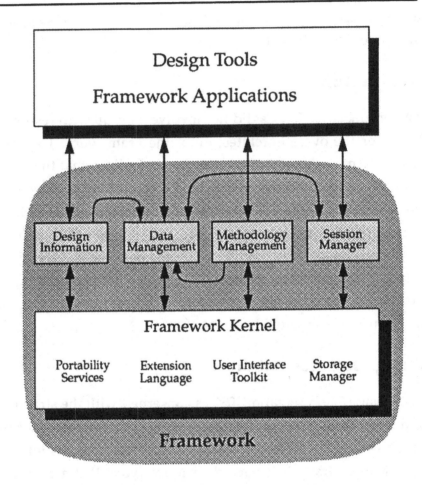

Figure 10.2: CFI Reference Architecture

The first version of the Design Representation Programming Interface was used at the DAC-90 demonstration, and a second version, which added bundles to the scalar capabilities of the first version, was used in the DAC-91 demonstration. This second version will be part of the *CFI 1.0* standard.

Design Methodology Management

Because of the lack of industry-wide agreement about how to do methodology management, the DMM group has focussed upon a set of file formats which can be used to statically describe the character of tool encapsulations, and tool run logs. The *Tool Encapsulation Standard* is a useful contribution to framework standardization, because it specifies a consistent and re-usable representation for declarative information concerning the encapsulation of external tools, without specifying any details of the implementation of the tool or the encapsulation itself. The *Tool Execution Log* format is used as an archival record of tool execution: it may be written by a framework or by a tool, and it provides an audit trail which may be used by *post hoc* design analysis tools.

Inter-Tool Communication

Given the desire to build an environment from heterogeneous tools, a means of communication between them is critical. CFI has tackled this by specifying a protocol for inter-tool messaging, along with a message dictionary, the semantics of which define a language by which tools may communicate both procedural information and data.

The first output from the ITC TSC was used as the basis of the 1991 DAC demonstration. Messages sent between tools and frameworks provided by many different suppliers were illustrated.The specification for this technology will be part of the *CFI 1.0* standard.

Operating System Interface

CFI has chosen to limit its standards for system interfaces to the domain of the Unix operating system. So far, the Systems Environment TSC has considered the choice of standard C libraries, and the design of a standard error handling system for both tools and framework.

User Interface

There can be few areas of software design which elicit such strong responses as the design of user interfaces, from programmers and users alike. Unfortunately, the state of the art in objective user interface quality measurement is still poor, and subjective and political factors play a strong part in the standardization process.

There is broad agreement that user interface standards should be based upon the X Window System, and that OSF's Motif look-and-feel should be the basis of the standard. Sun Microsystems has strongly argued that their user interface toolkit and style guide, - OpenLook™ - should also be part of the standard. This issue is characteristic of the nature of the standardization process: vested interests must be recognized and handled in the negotiation process.

Apart from these issues, work is proceeding with standards proposals for the command language and for user interface generation and customization.

Extension Language

As discussed in the previous chapter, CFI's Extension Language Working Group has selected the Scheme Programming Language [REE86] as the basis for the CFI-conformant extension language, while also proposing that an alternative syntax be defined for use by non-programmers, or those less familiar with Lisp, from which Scheme is derived. The CFI extension language is expected to be a part of the CFI 2.0 set of standards.

Continuing work includes specifying the interface between C code and the extension language, and a variety of CAD-specific extensions.

Technology CAD and
Component Information Representation

These two groups are new to CFI. They herald a widening of CFI's scope from the ECAD domain. The motivation is that although these areas are somewhat outside the original domain of CFI, the problems encountered are similar to those to which CFI is primarily addressed, and so some synergy is anticipated.

10.7   CFI in the Future

In three years the CAD Framework Initiative membership has demonstrated an unprecedented level of cooperation between traditionally competitive CAD tool suppliers, semiconductor and system houses, and hardware vendors. In addition, the technical progress to date has been very encouraging, with demonstrable standards for design representation, tool encapsulation and inter-tool communication. Interest in CFI continues high, and the plans for *CFI 2.0* are very ambitious.

The CAD Framework Initiative has provided not only a meeting ground for the development of standards, but also a hot-house within which the flower of CAD system standards may burgeon and grow with a rapidity little imagined by the initiators of the organization.

11 SUMMARY

In this book we have tried to describe the broad range of issues facing a CAD Framework designer. The past experiences of those who have tried to solve this problem is an invaluable guide to many of the trade-offs that must be made. However, we also believe that there is no complete CAD Framework and that there never will be. Technologies and understanding of the engineering design problems are changing far too rapidly for any system to meet all of the user and CAD tool needs for very long. It is important that this fact be considered as paramount when planning and developing a CAD Framework. The successful examples of similar technologies from the past are ones that were designed to evolve.

The development and use of standards, at all levels of the system, is important if we are to leverage all of the industries - computers, CAD tools, graphics hardware, software development, etc. - that play a key role in CAD system development. But standards can be a double-edged sword. They must be developed through, or from, use or they will inevitably not be able to find the appropriate compromise among the hundreds or thousands of competing factors which determine success.

11.1   Acknowledgments

We thank the members of the framework group at the Microelectronics and Computers Corp., in particular Paul Painter, Sandy Cavalli, Noel Strader and Laurence Brevard, for many helpful discussions. We also thank Bob Broderson, Wendell Baker, Paul Cohen, Rajiv Jain, Randy Katz, Peter Moore, Jan Rabaey, and Mario Silva for helping us to understand many of the issues presented in this book, as well as the needs of CAD designers. Our work in the area of CAD Frameworks has been supported since it began by DARPA and we would particularly like to thank Maj. John Toole for his encouragement and support. We also thank the Digital Equipment Corporation, Hewlett-Packard, the Microelectronics and Computers Corp., and the Semiconductor Research Corp. for their ongoing support. Finally, we would like to express our gratitude to Cadence Design Systems and Objectivity for providing time and resources to pursue the project in its final stages.

REFERENCES

[ALL90]   W. Allen, D. Rosenthal, K. Fiduk, "Distributed Method-
          ology Management for Design-in-the-Large",
          Proceedings of ICCAD-90, pp. 346-349, November
          1990.

[ALL91]   W. Allen, D. Rosenthal and K. Fiduk, "The MCC CAD
          Framework Methodology Management System",
          Proceedings of the 28th ACM/IEEE Design Automa-
          tion Conference, pp. 694-698, June 1991.

[AND89]   P. Anderson and L. Philipson, "Movie - An Interactive
          Environment for Silicon Compilation Tools", IEEE
          Transactions in CAD of IC's and Systems, Vol 8, No. 6,
          pp. 693-701, June 1989.

[ANS85]   Americal National Standards Institute, "American
           National Standard for Information Systems Computer
           graphics -- Graphical Kernel System (GKS) Functional
           Description", New York, NY, 1985.

[APP85]   *Inside the Macintosh*, Addison-Wesley, Reading, Massa-
           chusetts, 1985.

[AST76]   M. M. Astrahan *et al*, "System R: A relational approach
           to data management," ACM Transactions on Database
           Systems, Volume 1, pp. 97-137, 1976.

[BAR88]   T. Barnes, "CODEM: The Cooperative Development
           Model", unpublished paper, National Semiconductor
           Corporation, 1988 .

[BAR90]   T. Barnes, "SKILL^{TM}: A CAD System Extension
           Language", Proceedings of the 27th ACM/I'EEE
           Design Automation Conference, pp. 266-271, June 1990.

[BEN82]   J. Bennett, "A Database Management System for
           Design Engineers", Proceedings of the 19th ACM/IEEE
           Design Automation Conference, pp. 268-273, June 1982.

[BEN80]   J. Bentley, D. Haken, and R. Hon, "Fast Geometric
          Algorithms for VLSI Tasks", Proceedings of the IEEE
          Compcon, pp. 88-92, Spring 1980.

[BIL83]   G. Billingsley, "Program Reference for KIC", Report No.
          UCB/ERL M83/62, Electronics Research Laboratory,
          University of California at Berkeley, 1983.

[BOB81]   D. Bobrow and M. Stefik, "The LOOPS Manual", Tech-
          nical Report No. KB-VLSI- 81-13, Knowledge Systems
          Area, Xerox Palo Alto Research Center, 1981.

[BRE88]   M. Breuer, et. al., "Cbase 1.0: A CAD Database for VLSI
          Circuits Using Object Oriented Techniques", Proceed-
          ings of the IEEE ICCAD-88, pp 392-395, November
          1988.

[BRE90]   F. Bretschneider et al, "Knowledge Based Design Flow
          Management", Proceedings of ICCAD-90, pp. 350-353,
          November 1990.

[BRO75]   F. Brooks, Jr., *The Mythical Man-Month*, Addison-Wesley,
          Reading, Mass., 1975

[BRO85]   M. Brown, "Understanding PHIGS: The Hierarchical Computer Graphics Standard", Proceedings of Template, San Diego, California, 1985.

[BRO87]   J. Brouwers and M. Gray, "Integrating the Electronic Design Process", VLSI Systems Design, June 1987.

[BUS85]   M. Bushnell and S. Director, "Ulysses -- An Expert-System Based VLSI Design Environment", Proceedings of ISCAS-85, pp 893-896, 1985.

[BUS86]   M. Bushnell and S. Director, "VLSI CAD Tool Integration using the ULYSSES Environment", Proceedings of the 23rd ACM/IEEE Design Automation Conference, pp. 55-61, June 1986.

[BUS87]   M. Bushnell, "Ulysses -- An Expert-System Based VLSI Design Environment", Ph. D. Disseration, Department of Electrical Engineering, Carnegie-Mellon University, Pittsburgh, 1987.

[BUS89]   M. Bushnell and S. Director, "Automated Design Tool Execution in the Ulysses Design Environment", IEEE Transactions on CAD for IC's and Systems, Vol. 8, No. 1, pp. 279- 287, March 1989.

[CAS90]   A. Casotto, A. R. Newton, and A. Sangiovanni-Vincen-
          telli, "Design management based on design traces",
          Proceedings of the 27th ACM/IEEE Design Automa-
          tion Conference, pp. 136-141, June 1990.

[CFI91]   T. J. Barnes, editor, "Extension Language: Core
          Language Selection", CAD Framework Initiative Docu-
          ment # ARCH-91-G-1, May 1991.

[CHE76]   P.P.-S, Chen, "The entity-relationship model: Toward a
          unified view of data", ACM Transactions on Database
          Systems, Vol 1, pp. 9-37, March 1976.

[CHE80]   P. P.-S. Chen, editor, *Entity-Relationship Approach to
          Systems Analysis and Design*, North-Holland,
          Amsterdam 1980.

[CHE88]   G. Chen and T. Parng, "A Database Management
          System For A VLSI Design System", Proceedings of the
          25th ACM/IEEE Design Automation Conference, pp.
          257-262, June 1988.

[CHO88]   H. Chou and W. Kim, "Versions and Change Notifica-
          tion in an Object-Oriented Database System",
          Proceedings of the 25th ACM/IEEE Design Automa-
          tion Conference, pp. 275- 281, June 1988.

[CHU83]   K. Chu, et. al., "VDD - A VLSI Design Database
           System" , ACM SIGMOD Conference on Engineering
           Design Applications, 1983.

[DAN87]   J. Daniell, A. Dewey and S. Director, "Artificial Intelli-
           gence Techniques: Expanding VLSI Design Automation
           Technology", Research Report No. CMUCAD- 87-38,
           SRC-CMU Research Center for Computer-Aided
           Design, Carnegie-Mellon University.

[DAN89]   J. Daniell, S. Director, "An Object Oriented Approach to
           CAD Tool Control Within a Design Framework",
           Research Report No. CMUCAD-89-15, SRC-CMU
           Research Center for Computer-Aided Design, March
           1989.

[DEC89]   *Guide to the XUI User Interface Language Compiler*, Digital
           Equipment Corporation, 1989.

[DEM87]   L. Demers, P. Jacques, S. Fauvel and E. Cerny,
           "CHESHIRE: An Object-Oriented Integration of VLSI
           CAD Tools", Proceedings of the 24th ACM/IEEE
           Design Automation Conference, pp 750-756, June 1887.

[DEW86]   P. Dewilde, Editor, *Data Management for Hierarchical and Multiview VLSI Design*, Delft University Press, Delft, The Netherlands, 1986.

[EBE89]   C. Ebeling and Z. Wu, "WireLisp: Combining Graphics and Procedures in a Circuit Specification Language", Proceedings of the 1989 IEEE International Conference on Computer-Aided Design, pp. 322-325, November, 1989.

[ECK88]   D. Ecklund and F. Tonge, "A Context Mechanism to Control Sharing in a Design Database", Proceedings of the 25th ACM/IEEE Design Automation Conference, pp. 344- 350, June 1988.

[EIA87]   EDIF Steering Committee, *EDIF Electronic Design Interchange Format Version 2 0 0*, Electronic Industries Association, 1987.

[FEL79]   S. I. Feldman, "Make - A program for maintaining computer programs", *UNIX Programmers Manual*, Bell Laboratories, Murray Hill, New Jersey, 1979.

[FLA87]   B. Fladung, *The XLISP Primer*, Prentice-Hall, Englewood Cliffs, New Jersey, 1987.

[FOL82]   J. Foley and A. van Dam, *Fundamentals of Interactive Computer Graphics*, Addison-Wesley, Reading Mass., 1982.

[FOL84]   J. Foley, Victor Wallace and Peggy Chan, "The Human Factors of Computer Graphics Interaction Techniques", IEEE Computer Graphics and Applications, pp 13-46, November 1984.

[GOO83]   J. R. Goodman, "Using cache memory to reduce processor-memory traffic", International Symposium on Computer Architecture, pp. 124-131, June 1983.

[GOS86]   J. Gosling, "SunDew - A Distributed and Extensible Window System", Methodology of Window Management (Proceedings of an Alvey Workshop at Cosener's House), Springer-Verlag, 1986.

[GOT87]   K Gottheil, et. al., "The Cadlab Workstation CWS - An Open, Generic System for Tool Integration", Cadlab Report 3/87, Padderborn, Germany, December 1987.

[GUP89]   R. Gupta, et. al., "An Object-Oriented VLSI CAD Framework: A Case Study in Rapid Prototyping", IEEE Computer, Vol. 22, No. 5, pp. 28-37, May 1989.

[GUT84]   A. Guttman, "New Features for a Relational Database
          System to Support Computer Aided Design", Ph. D.
          Dissertation, Department of Electrical Engineering and
          Computer Sciences, Univesity of California, Berkeley,
          1984.

[HAR86]   D. Harrison, et. al., "Data Management and Graphics
          Editing in the Berkeley Design Environment," Proceed-
          ings of the IEEE ICCAD-86, pp. 20-24, November 1986.

[HEL75]   G. Held, et. al., INGRES - A Relational Data Base
          System, Proceedings of the AFIPS, Vol. 44, pp. 409-416,
          1975.

[HOR67]   G. Hornbuckle, "The Computer Graphics/User Inter-
          face", IEEE Transactions on Human Factors in
          Electronics, Vol. 8, No. 1, pp. 17-22, March 1967.

[INT85]   "VHDL Language Reference Manual, Version 7.2",
          Technical Report No. IR- MD-045-2, Intermetrics, Inc.,
          Bethesda, MD, August 1985.

[JOH89]   N. Johnson, *AutoCAD: The Complete Reference*, Osbourne
          McGraw/Hill, Berkeley, CA, 1989.

[JUL86]   C. Jullien, et. al., "A Database Interface from an Integrated CAD System", Proceedings of the 23rd ACM/
IEEE Design Automation Conference, pp. 760-767, June
1986.

[KAH87]   H. Kahn; *SIDESMAN: A CAD System for VLSI Design"*,
*Intelligent CAD Systems 1: Theoretical and Methodological
Aspects*, North-Holland, 1987.

[KAH87b]   H. Kahn; "Do Engineers REALLY need Artificial Intelligence", IEEE International Workshop on AI --
Applications to CAD-Systems for Electronics, October
1987.

[KAT85]   R. Katz, *Information Management for Engineering Design*,
Springer- Verlag, 1985.

[KAT86]   R. Katz, et. al., "A Version Server for Computer-Aided
Design Data", Proceedings of the 23rd ACM/IEEE
Design Automation Conference, pp. 27-33, June 1986.

[KEL84]   K. Keller, "An Electronic Circuit Framework", Report
No. UCB/ERL M84/54, Electronics Research Laboratory, University of California at Berkeley, 1984.

[KER87]   B. Kernighan and D. Ritchie; *The C Programming Language*, Prentice-Hall, Englewood Cliffs, NJ, 1978.

[KRU90]   J. Krueger, "Experiences with an Extension Language for Engineering Information Systems", unpublished paper, Honeywell Systems and Research Center, Minneapolis, Minnesota.

[LAI86]   L Lai and G. Wood, "SKILL - An Interactive Procedural Design Environment", Proceedings of the CICC '86, pp. 544-547.

[LIC62]   J. Licklider, "On-line Man-machine Communication", Proceedings of the Spring Joint Computer Conference, Vol. 21, pp. 113-128, 1962.

[LIN86a]   J. L. Linn and R. I. Winner editors, *The Department of Defense Requirements for Engineering Information Systems: Volume 1*, The Institute for Defense Analysis, Alexandria, Virginia, 1986.

[LIN86b]   J. L. Linn and R. I. Winner editors, *The Department of Defense Requirements for Engineering Information Systems: Volume 2*, The Institute for Defense Analysis, Alexandria, Virginia, 1986.

[MCC88]   J. McCormack, P. Asente and R. Swick, "X Toolkit Intrinsics - C Language Interface", Massachusetts Institute of Technology, 1988.

[MCC89]   "C Module Editor: Version 2.0 User's Guide", MCC Technical Report No. CAD- 043-89, Microelectronics and Computer Corporation, 1989.

[MEY88]   B. Meyer, *Object-Oriented Software Construction*, Prentice-Hall, New York, NY, 1988.

[MIL89]   J. Miller *et al*, "The object-oriented integration methodology of the Cadlab work station design environment", Proceedings of the 26th ACM/IEEE Design Automation Conference, Las Vegas, Nevada, pp. 807-810, 1989.

[MIT85]   T. Mitchell, L. Steinberg and J. Shulman, "A Knowledge-Based Approach to Design", LCSR-TR-65, Department of Computer Science and Laboratory for Computer Science Research, Rutgers University, January 1985.

[MOO86]   D. Moon, "Object-Oriented Programming with Flavors", Proceedings of the ACM OOPSLA-86, pp. 1-8, 1986.

[NEW73] W. Newmann and R. Sproull *Principles of Interactive Computer Graphics*, McGraw/Hill, 1973.

[NEW81] R. Newton *et al*, "Design Aids for VLSI: The Berkeley Perspective", IEEE Transactions on Circuits and Systems, vol CAS-28. pp. 668-680, July 1981.

[OUS81] J. Ousterhout, "Caesar: An Interactive Editor for VLSI Layouts", VLSI Design, Vol. 2, No. 4, Fourth Quarter 1981.

[OUS84a] J. Ousterhout, "Corner Stitching: A Data-Structuring Technique for VLSI Layout Tools," IEEE Transactions on CAD for IC's and Systems, Vol. 3, No. 1, pp. 87-100, January 1984.

[OUS84b] J. Ousterhout, et. al., "Magic: A VLSI Layout System", Proceedings of the 21st ACM/IEEE Design Automation Conference, pp. 152-159, June 1984.

[PAS89] W. Paseman, "Tools on a New Level", Unix World, pp. 69-77, June 1989.

PIL89] R. Piloty *et al*, "IREEN(V3.13) - Data base concepts and representation of VHDL design data", Technische

Hochschule Darmstadt No. RO 89/3, Darmstadt, Germany, June 1989.

[REE86] J. Rees *et al*, "Revised[3] report on the algorithmic language scheme", ACM Sigplan Notices, vol 21 No. 12, December 1986.

[ROB81] Roberts, et. al., "A Vertically Organized Computer-Aided Design Database", Proceedings of the 18th ACM/IEEE Design Automation Conference, pp. 595-602, June 1981.

[ROG85] C. Rogers, et. al., "MCNC's Vertically Integrated Symbolic Design System", Proceedings of the 22nd ACM/IEEE Design Automation Conference, pp. 62-68, June 1985.

[ROS85] J. Rosenberg, "Geographical Data Structures Compared: A Study of Data Structures Supporting Region Query", IEEE Transactions on CAD for IC's and Systems, Vol. 4, No. 1, pp. 53-67, January 1985.

[RUB87] S. Rubin, *Computer Aids for VLSI Design*, Addison-Wesley, 1987.

[SCH86]   R. Scheifler and J. Gettys, "The X Window System",
          ACM Transactions on Graphics, Vol. 5, No. 2, pp. 79-
          109, April 1986.

[SCH88]   R. Scheifler, J. Gettys and R. Newman, *X Window
          System: C Library and Protocol Reference*, Digital Press,
          Bedford, Mass., 1988.

[SID80]   T. Sidle "Weakness of Commercial Database Manage-
          ment Systems in Engineering Design", Proceedings of
          the 17th ACM/IEEE Design Automation Conference,
          June 1980.

[SIE84]   D. Sieworek, et. al., "DEMETER Project: Phase 1
          (1984)", Research Report CMUCAD-84-35, SRC-CMU
          Center for Computer-Aided Design, Department of
          Electrical and Computer Engineering, Carnegie-Mellon
          University, July 1984.

[SIL89]   M. Silva, et. al., "Protection and Versioning for the OCT
          Environment", Proceedings of the 26th ACM/IEEE
          Design Automation Conference, pp. 264-262, June 1989.

[SMI75]   C. Smith, "Calma's GPL^{TM} Language - A Program
          Language for Custom, Turnkey Graphic Systems",
          Proceedings of the Fall 1975 IEEE Compcon, 1975.

[SMI82]   D. Smith, et. al., "Designing the Star User Interface",
           Byte, pp. 242-282, April 1982.

[SMI88]   W. Smith, et. al., "Flexible Module Generation in the
           FACE Design Environment", Proceedings of the IEEE
           ICCAD-88, pp. 396-399, November 1988.

[STA87]   R. Stallman, GNU Emacs Manual, Sixth Edition, Emacs
           Version 18, Free Software Foundation, March 1987.

[STE75]   G. L. Steele and G.J. Sussman, "Scheme: An Interpreter
           for the Extended Lambda Calculus", Memo 349, MIT
           Artificial Intelligence Laboratory, 1975.

[STE90]   Guy L. Steele, Common Lisp: The Language, 2nd Edition,
           Digital Press, Bedford, MA, 1990.

[STE84]   L. Steinberg and T. Mitchell, "A Knowledge Based
           Approach to VLSI CAD: The REDESIGN System",
           Proceedings 21st ACM/IEEE Design Automation
           Conference, pp. 412- 418, June 1984.

[STR87]   B. Stroustrup, The C++ Programming Language,
           Addison-Westley Publishing, Reading, MA., July, 1987.

[SUT63]   I. Sutherland, "Sketchpad - A Man-machine Graphical
           Communication System", Proceedings IFIPS Spring
           Joint Computer Conference, pp. 329-345, 1963.

[SWA88]   G. Swan, et. al., "Design Management in a Workstation
           Environment", Proceedings of the 22nd Annual Hawaii
           International Conference in the System Sciences,
           January 1988.

[TES81]   L. Tesler, "The Smalltalk Environment", Byte, pp. 90-
           147, August 1981.

[TIC82]   W. Tichy, "Design, Implementation, and Evaluation of a
           Revision Control System", Proceedings of the 6th IEEE
           International Conference on Software Engineering,
           Tokyo, September 1982.

[ULL82]   J. Ullman, *Principles of Database Systems*, Computer
           Science Press, Rockville, MA, 1980

[WEI86]   S. Weiss, et. al., "DOSS: A Storage System for Design
           Data", Proceedings of the 23rd ACM/IEEE Design
           Automation Conference, pp. 41-47, June 86.

[WID88] J. Widya, et. al., "Concurrency Control in a VLSI Design Database", Proceedings of the 25th ACM/IEEE Design Automation Conference, pp. 357-362, June 1988.

[WOL88] P. v. d. Wolf and T. G. R. v. Leuken, "Object type oriented data modeling for VLSI data management", Proceedings of the 25th ACM/IEEE Design Automation Conference, Anaheim, California, pp. 351-356, June 1988.

[WOL89] W. Wolf, "How to Build a Hardware Description and Measurement System on an Object-Oriented Programming Language", IEEE Transactions on CAD of IC's and Systems, Vol 8, No. 3, pp. 288-301, March 1989.

[WON79] S. Wong and W. Bristo, "A Computer Aided Design Database" Proceedings of the 16th ACM/IEEE Design Automation Conference, June 1979.

[ZIP85] R. Zippel and C. Clark, "Schema: An Architecture for Knowledge Based CAD", Research Report VLSI Memo No. 85-271, Department of Electrical Engineering and Computer Science, Mass. Institute of Technology, October 1985.

INDEX

A

C

D

E

F

G

H

I

K

L

M

N

O

P

R

S

T

U

V

W

X